Airbus vs. Boeing: Strategy Wars, Tactical Dogfights, High-G Maneuvers & the Photo Finishes –

Single Abridged Edition

RAJAT NARANG

Airbus vs. Boeing: Strategy Wars, Tactical Dogfights, High-G
Maneuvers & the Photo Finishes – 1970s to 2020

Front Cover Image: John Dubois, Pexels.com, Image ID:
11020029

Cover Photo by Tom Dubois from Pexels

Back Cover Image: Quintin Gellar, Pexels.com, Image ID:
6604555

Published by: 9000RPM Publishing Works

Cover Page Design: Canva.com

Airbus vs. Boeing: Strategy Wars, Tactical Dogfights, High-G
Maneuvers & the Photo Finishes – 1970s to 2020

DEDICATION

Dedicated to the 'Supreme One' up above for the 'Vision & Drive' and the 'two unshakable pillars of my life' for being an indispensable part of this journey. The first being my wife, Bhumika, who as my life's co-pilot and navigator, has always provided the crucial inspiration & played a critical 'role' in charting & paving this least travelled path over the past almost two decades sharing the cockpit on numerous sorties & missions through the life's danger zones facing severe turbulences & weathering unrelenting storms on multiple occasions in the process invariably while the second has stood tall with us as the virtual ATC tower with unwavering & unconditional support throughout as the eternal 'Archangel of Hope.'

Due & invaluable credits to the Creator for providing the critical propulsive thrust, maneuverability and the tenacity to be able to somehow continue & strive on the assigned flight path amid phases of tremendous head winds and turbulence. Also, sincere & heartfelt gratitude to the 'two towers of strength & compassion', who with their deep professional competence & experience along with superb teams of on-field commanders, provided the much needed reverse thrusters just in time to abort my almost unplanned take-off to the other world due to a sudden physical health breakdown a couple of years back and put things again in working order post a major overhaul through a complex neurosurgical intervention followed by rehab for the return to action after a year-long hiatus…That bond, forged in titanium, remains special even to this day!

Stay Blessed!

CONTENTS

Disclaimer

The Author does not have any kind of financial investments, direct business associations or financial stakes of any sort in any of the companies included & mentioned in this work. It does not promote or endorse any particular company or industry player & its products over others in any manner whatsoever. The overall scope & directional reference compass of this work primarily is the global narrow body aircraft market with focus on relative, comparative analysis from a longitudinal strategy perspective under the purview of the prevailing strategic & operating context.

The analysis has been purely anchored on factual evidences available in the public domain based information sources and utmost effort has been made to ensure fair play to present the most objective view of reality against the backdrop & in light of the available facts rather than taking a biased view skewed towards any of the two industry behemoths covered. The approach taken, thus, is intended to clearly avoid the scope for straying towards the dark zone marked by subterfuge & sabotage tactics, black ops, undercover missions, propaganda and/or shows of mudslinging.

The opinions expressed throughout are purely author's personal and the views & judgments presented are neither directed or targeted at anyone nor meant as pontification of any sort and are solely based on objective assessment of prevailing situations, decisions & outcomes. The names of products, systems and/or brand names mentioned wherever through the work are intellectual property of their respective owners and their mention has been done only & purely for information & creative purposes and it does not indicate or reflect and should not be construed as any kind of promotion or endorsement of any kind. Airbus, which has been renamed couple of times since coming into existence, has been referred to simply as

Airbus; the current company name since 2014; throughout for simplicity, relevance & consistency reasons with reference to present context & current reality. Also, there is no intent to glorify warfare over other humane methods of settlement of disputes. The objective is to explore and extrapolate across domains with the activity anchored firmly on the unshakable common ground between warfare & business which is strategy. The lenses used to view and analyze, therefore, have primarily been of strategy rather than history which has invariably led to limiting the scope for deep dives to areas & angles relevant from the strategy perspective. It is also not intended as a technology/technical guide with the inclusion of some technical aspects & performance parameters for analysis done purely from a relative strategy perspective.

The information and facts contained herein are believed to be correct at the time of publication but cannot be guaranteed. All the information presented has been derived from reliable sources, reasonably verified & has been presented purely & solely for informational purposes only. The views expressed throughout are based on broad analysis & assessments only and thus should not be substituted for professional advice & opinion of any kind prior to decision-making. The author expressly disclaims any and all liability to any person or entity pertaining to potential outcomes or consequences of any decisions or actions taken based on the contents of this publication.

The analysis presented includes & cites quotes, comments, statements and very short excerpts from public reports; derived from public domain based information sources; originating from & pertaining to industry leaders, senior company executives, industry & company analysts, journalists, aviation authorities, industry bodies & other authority figures from the commercial aviation industry with reference to particular situational contexts with proper attribution & due credits provided under fair terms of use.

The incorporation of these has been done throughout purely to provide readers with actual prevailing viewpoints, opinions & industry stakeholders' take on the specific operating and/or strategic context being analyzed as contextual evidences for the lines of analysis presented, angles of view taken and positions adopted on strategy moves as well as inferences & conclusions drawn. The approach has also been taken for the readers to be able to fairly construct the contours of the subject matter clearly, review them, be able to form their own unbiased judgments, opinions & viewpoints and be able to explore the subject further by treading along the provided pathway & digging deeper. For any issues or concerns with the same please feel free to contact.

ACKNOWLEDGMENTS

Sincere thanks, due regards and special acknowledgements to all the well known as well as unsung aviation engineers, airplane designers, test pilots & pilots, adventurers, entrepreneurs, industry personnel, enthusiasts and aerospace journalists, analysts as well as publishers who have dedicated their lives to shaping the industry's evolution through time, those doing it at the moment and the ones yet to come to take on the mantle and steer this wonderful journey forward!

Airbus vs. Boeing: Strategy Wars, Tactical Dogfights, High-G Maneuvers & the Photo Finishes – 1970s to 2020

Chapter 1

THE 'TWO-TEAM' LEAGUE AND A RELENTLESS ROOKIE ON THE RADAR

"He will win who knows when to fight and when not to fight" – Sun Tzu, The Art of War

"We will not do with Bombardier what Boeing did with Airbus" [1] said John Leahy, Chief Operations Officer - Customers at Airbus in December 2010 formally announcing Airbus' declaration of war against the Bombardier's C-Series Program.

"Airbus will not allow Bombardier to establish a beachhead with the C-Series against the A319. A re-engined A320 family would destroy the business case for the C-Series" [1] bellowed the raging deputy, Tom Williams, EVP Programs at Airbus.

The setting was the company's investor day at Airbus headquarters in December 2010 when Airbus announced that it was going to take on Bombardier & its latest C-Series program head-on to crush it as amply reflected in the charged up statements of the two Airbus senior commanders (above). This had followed Bombardier's winning of a major aircraft deal from Frontier Airlines, owned by Republic Airways Holdings; a staunch Airbus loyalist operating a pure Airbus aircraft fleet comprising A319/A320 aircrafts. The Airbus War Room had been abuzz and on high alert with interceptor aircrafts having already been scrambled, battle plans being chalked out and air defense batteries in full readiness mode for action post this incursion of a rather well known bogie which had simply sneaked into its airspace and had been tracking & locking on to strategic targets with utter impunity.

This audacious predation on its territory with a coup had enraged Airbus which treated it as Bombardier's crossing the line in its bid to foray into the Airbus stronghold and something which had to be checked and replied to in full measure. John Leahy, then COO-Customers for Airbus, explicitly mentioned this intent while referring to Bombardier. He declared vociferously in his war cry that Airbus will not ignore Bombardier and the attempted invasion in its nascent stages and will rather nip it in the bud, unlike Boeing, which had ignored the emergence of Airbus in the early stages in the 1970s & 80s, which in fact, has turned out to be the ultimate transatlantic arch-rivalry & duopoly in the global commercial aircraft market.

This was the stage at which the die was ultimately cast for the launch of the re-engined A320neo by Airbus; which it had been deferring since first announcing the program in February 2010 at the Singapore Airshow 2010; wherein the company had announced that it will be formally launching the same at the Farnborough Airshow in July 2010. The target & the mission objective was absolutely clear, to take down the invader, the Bombardier's C-Series before it could do any more damage.

.

Both Airbus and Boeing had been taking the emerging threat from the incoming bogie, C-Series, very seriously treating it as a potential intrusion into their respective airspaces. Boeing, however, chose to ignore Bombardier initially but later overreacted when the C-Series started poaching onto Boeing's territory competing with the lower end 737 variants for orders from airline customers. Airbus, on the contrary, had been circumspect, having adopted a much more cautious approach given its business' much higher degree of alignment with the narrow body aircraft market spearheaded by the fledgling A320 aircraft family comprising of the quartet of A318, A319, A320 & the A321.

The A319 has been the Airbus' campaigner in the 130-140 seat aircraft segment with it essentially being a shrunk variant of the A320 with a reduced overall length and lesser passenger carrying capacity. Airbus had launched the A319 in 1992; powered by the de-rated variants of CFM56 & V2500 engines as options; at a projected program development cost of EUR 250 million with ILFC as the launch customer.

The A319 had been followed by the development of the A318, the smallest variant of the A320 aircraft family. Airbus first talked about the A318 in 1998 with the European aerospace giant actively locking on to the 100 seat aircraft segment. The A318, with a seating capacity of 107 passengers, was formally announced at the Farnborough Airshow in 1998.

The A319 had its maiden flight in 1995 & entered service in 1996 post certification. The smaller sibling, the A318, made it to the skies in early 2002 followed by entry into service in 2003. This hectic Airbus activity also forced arch-rival Boeing to launch its significantly revamped & re-engined 737 Next Generation or NG aircraft series in the early 1990s intended at effectively taking on a growing Airbus clout. The 737NG series comprised the 737-600, -700, -800 & -900 variants powered by the CFM56-7 engines with the first one, the 737-700, getting airborne with Southwest in late 1997.

This was precisely the time when Bombardier, buoyed by the Airbus & Boeing product strategies; especially, the launch of the smaller 100+ seat aircraft variants; had been brewing up plans of its own to foray into this 100-140 seat aircraft market segment. Airbus & Boeing had already tapped into the segment by the late 1990s with the launch of the shrunk members of their A320 & 737 aircraft programs taking a low-cost, low-risk derivative approach. However, the key difference between Airbus & Boeing and Bombardier, in terms of approach, was that Bombardier's plan was exorbitant in nature and was based on a clean sheet aircraft design while the company was still an aviation minnow limited to the business & regional aircraft segments busy cranking out CRJs and Challenger as well as Global series business jets.

Airbus' Options & the Equation – Clean Sheet vs. Re-Engine

After almost two decades of the entry into service of its flagship narrow body aircraft, the A320; which had by now firmly entrenched itself in the global narrow body aircraft market; Airbus had been looking at further enhancing the overall efficiencies of the program in 2006. With this intent, Airbus had initiated the A320 Enhanced or A320E program in 2006 aimed at achieving further fuel efficiency gains through the proposed incorporation of large winglets (which were expected to bring in 2% gains), aerodynamic refinements targeting 1% improvement apart from potential weight savings and the incorporation of a new cabin with all these collectively targeting overall gains pegged in the 4% to 5% range.

The share of gains originating from the prevailing generation of CFM56 & the V2500, however, were limited and pegged at a mere 1% based on the CFM56 Tech Insertion and later the V2500 SelectOne improvement packages. The CFM's Tech Insertion package was launched by CFM International in 2004 and became available in 2007.

Targeted at the existing in-service CFM56-5B/7B engines, the package included an upgrade kit for the compressor, upgrades to the combustor and high as well as low pressure turbine hardware translating into a 1% improvement in fuel burn rate along with higher durability and a longer time on wing besides corresponding lower maintenance costs while complying with the ICAO's Committee of Aviation Environmental Protection Standards (CAEP/6) which had come into effect in 2008.[2] By 2009, Airbus had come full circle on the A320 program having completed the introduction of the sharklet based wingtip device which was delivering a 3.5% reduction in fuel burn rate and had also introduced a new reduced weight aircraft cabin.

At this stage, considering an expensive, clean sheet approach to A320 succession with fuel efficiency gains of just 4% to 5% at the most (without next generation engines) did not make any economic sense factoring in the huge development costs involved and given the fact that the next generation, high-bypass turbofan engines (which had already been under development at the time) were likely to become available within a decade's span and could have simply taken the overall fuel efficiency gains to the almost 15% level. "Who's going to roll over a fleet to a new generation aircraft for 5% better than an A320 today? Especially if another 10% improvement might be coming in the second half of the next decade based on new engine technology"[3] said John Leahy, the then sales chief of Airbus commercial aircrafts. Re-engining, thus, would have been the most logical course of action for Airbus going forward given the end of road in sight for improvement initiatives on the A320 coupled with the rapid evolution of market landscape with the emergence of a serious threat from the C-Series which had just been launched by Bombardier at the 2008 Farnborough Airshow. The C-Series, in fact, had already started competing with & eating into Airbus' market shares by targeting & eyeing key airlines operators and even staunch Airbus loyalists.

The decision to launch re-engined variants of their respective existing narrow bodies was not going to be an easy one either for Airbus or for Boeing as they both had to thoroughly evaluate the business case for the same. The aerospace giants were fearful that any potential increase in fuel efficiency gains from re-engining might get negated, to a large extent, by the likely, commensurate increase in additional capital costs of procurement along with an increase in maintenance costs and requisite spares inventory which collectively could have destroyed the overall business case for re-engining.

Bombardier, on the contrary, had been extremely confident of its assessment that the C-Series, as a radical & clean sheet design powered by next generation engines, would have a direct cost advantage over re-engined Airbus' A320 or Boeing's 737 aircrafts. Andy Shankland, Vice President, Marketing at Airbus said while speaking at the 24th Annual Aircraft Finance and Commercial Aviation Conference held in March 2010 quoting John Leahy, (who had spoken a week earlier at the ISTAT conference about the prospects of A320neo) that "it would be nice to have an announcement at

the Farnborough Airshow 2010"[4]. Boeing's Randy Tinseth said "Boeing
was in "no hurry" to make a decision, while one would be forthcoming this
year. [4]"

John Leahy had also said at the Singapore Airshow earlier in February
2010: "Airbus is seeing a surprising amount of interest from customers in
its re-engining proposals. Some of this might have been started by
Bombardier with its GTF powered C-Series, claiming to be 15% better than
an A320. But if we do this (re-engning), you can have your A320 and still
have your 15% lower fuel burn[5]".

John Leahy also made another key remark at the Singapore Airshow
2010 which was significant. He said "The timing of the re-engined A320
family means that the earliest an all-new single aisle airliner will come from
Airbus, dubbed "A30X", is in the middle of the next decade, around 2024-
25"[5]. Looking at the prevailing commercial aviation market scenario prior
to COVID-19 outbreak, this brand new A320 replacement was slated to
come around sometimes in late 2020s or early 2030s. However, in a post
COVID-19 world in 2020, this interim Airbus decision is likely to be
skipped entirely owing to current market realities and the A320 program is
all set to leapfrog technologically by the mid-2030s now, as per latest Airbus
plans.

"Bombardier's C-Series has Airbus and Boeing 'spooked'. The CS100
and CS300, powered by Pratt & Whitney's PW1000G Geared Turbo Fan
(GTF), is directly competitive with the Airbus A318/319 and the Boeing
737-700. The two larger OEMs are considering the GTF, CFM's Leap-X
and the Rolls-Royce 2- and 3-stage engines for the A320 and 737 families"
said David Swan of RBS Aviation Capital, a panelist at the very same

conference[4]. This was way back a decade ago in 2010 when both Airbus & Boeing were virtually at wits end to make a quick but well calculated "Go, No-Go" decision over the launch of their respective re-engining programs amid the industry at large having already started the countdown for the same.

No such announcement came forth from either Airbus or Boeing at the Farnborough Airshow 2010 contrary to the industry expectations which surprised many. However, Airbus, of the duo, was the first one to blink on this with the launch decision for the A320neo announced by the company on December 01, 2010 in an unusual setting in a very low profile manner which was quite unusual & uncharacteristic of the European aerospace giant. FlighGlobal reported & commented on the same, "The Airbus decision comes after a protracted 'will they, won't they' debate over hanging new engines on the company's best-selling models. Boeing still has yet to announce its future plans on its 737"[6] capturing & expressing the prevailing market sentiment quite succinctly.

The 'Decade of the Re-Engined Aircrafts' which followed the 'Decade of the Clean Sheet Aircrafts'

The market tide had turned with the launch of the Airbus A320neo heralding the onset of the decade of the re-engined aircrafts; as it has truly turned out to be; witnessing the launch of 4 re-engined commercial aircraft programs led by the Airbus' A320neo aircraft family, Boeing's 737 MAX, the fourth generation of the 737 program, Embraer's E-Jets E2 regional aircraft family and the Boeing's 777X wide body aircraft program.

This was in complete contrast to the first decade of the 21st century which had belonged completely to new, clean sheet aircrafts with a herd of new aircraft programs launched by the industry spearheaded by the

Boeing's 787 & Airbus' A350XWB wide bodies, Bombardier's C-Series, Mitsubishi's MRJ regional aircraft program, COMAC's ARJ21 regional & C919 narrow body aircraft programs, Sukhoi's SSJ100 regional jet and the Russian MC-21 narrow body aircraft program.

This market rally towards new, clean sheet aircraft programs, led by new market entrants, was propelled by the advent of Pratt & Whitney's next generation Geared Turbofan (GTF) engines for single aisle aircrafts with most of the new narrow body & regional aircraft program launches featuring the GTF at least as an engine option. However, with the exception of old masters of the craft, Airbus & Boeing, along with the crusader Bombardier & the ARJ-21 to some extent; all of these new aircraft programs are still under different stages of development and are yet to fully enter commercial revenue service.

Coming or rather going back to the A320neo launch by Airbus; the ball was in the Boeing's court now and it had to take a call now on how to counter the opening salvo from arch rival Airbus in a strategic market segment for both the OEMs within a duopolistic market set up. Boeing was clearly on the reactive now, a usual & familiar place for the American giant in the narrow body market segment, as it had been used to operating in the reactive mode with a largely competition focused strategy orientation since the launch of the 737 Original in the mid-1960s. It is said that history has the weird proclivity of repeating itself and on this occasion it indeed was repeating itself in an uncannily similar way.

The market introduction of the Airbus' A320 aircraft family in the second half of the 1980s & its subsequent, broader market acceptance as a next generation European narrow body aircraft incorporating digital fly by wire controls (for the first time in an commercial airliner) had unnerved Boeing which ultimately led to the decision to launch the Boeing 737NG, the 737's third generational avatar.

At that time in the 1980s, Airbus' decision to incorporate fly by wire technology along with flight envelope protection in a commercial aircraft program coming from a rookie proved to be a real game changer and radically tilted the strategic calculus (in the narrow body aircraft market) from then on in favor of Airbus ultimately resulting in a European dark horse going on to almost win the narrow body race by the end of the second decade of the 21st century!

"Know your Enemy, Know his Sword". – Miyamoto Musashi, The Book of Five Rings

Chapter 2

A 'GAME CHANGER' BIRD AND A MAVERICK WITH THE VISION, ON A MISSION AND THE '$1 TRILLION+ DIFFERENCE' IT ULTIMATELY MADE…

"You are remembered for the rules you break": General Douglas MacArthur, U.S. Army

Let's hover back in time to that momentous decision making moment in the 1980s when the digital fly by control technology used on Dassault's Mirage 2000 fighter jets inspired Airbus engineers to test the same on the A300. For Airbus, it would have given the rookie a definite head start with a technological edge, competitive differentiation & a much enhanced value proposition (for customers) with which Airbus could have taken on a mighty competitor much more effectively on a terrain well known to the enemy under difficult operating conditions.

However, the decision was not an easy one for Airbus by any means. Roger Béteille, Airbus President and also regarded as one of the founding fathers of the Airbus consortium, had termed the decision to ultimately introduce fly by wire (FBW) technology along with flight envelope protection in the early 1980s as one of the most difficult decisions he had to ever make in his career. On this he had said: "Perhaps we were too bold, but we had no choice. Either we were going to be first with new technologies or we could not expect to be in the market[1]". As per Airbus, he is also credited for instilling the core value & culture of listening to customers while developing new products.

The key point was that the FBW technology was mature, proven & not under development and had been in use extensively across military combat jets of that generation. Additionally, there was some heritage & legacy involved as well. Airbus precursor, Sud Aviation, which later became Aerospatiale, had used analogue, computer-driven FBW controls a decade earlier on the joint Franco-British supersonic transport aircraft program, the Concorde, which had entered service in 1976.

Airbus had also used the electrical signaling system on the secondary flight control systems on the A300/A310 program, the world's first twinjet wide body aircrafts, in the early 1970s. So, Airbus did take a tough call & ultimately got it right with the A320 family entering service in April, 1988 and going on to propel Airbus' rise as a European aerospace powerhouse by the turn of the century while giving a harrowing time to the Boeing 737.

The most radical & revolutionary aspect of the usage of the FWB technology on the A320 in the 1980s was the incorporation of the side stick controls (passive) instead of the traditional central control column with the move geared towards reducing overall weight as well as pilot workload while enhancing overall aircraft safety through flight envelope protection.

The Airbus strategy of using very similar control mechanisms & characteristics for its family of FBW airliners, which have been launched since then, had another key added advantage for the airlines customers given commonality of cockpits which enabled same pilot type ratings to be maintained across aircrafts belonging to the same aircraft family. The usage of the common man-machine interfaces & control mechanisms across the A320 aircraft family led to acquisition of cross type/cross crew qualification for pilots across aircrafts belonging to the same family possible at a much faster pace and at minimal additional training costs for the airlines, thereby, enabling the pilots to transition to other aircrafts of the same family by fast tracking their qualification. This commonality philosophy has been maintained by Airbus even in the wide body aircrafts and continues to this day. For instance, Airbus claims that a pilot certified for the A330 can simply transition to the A350XWB in a matter of 8 days[16] without requiring full simulator sessions.

These identical cockpits across aircrafts, based on commonality approach, provide tremendous flexibility to airlines. Boeing, unlike Airbus, had not taken & pursued this kind of a synchronic, well planned approach to product strategy ab initio aimed at the development of an entire aircraft family from scratch leveraging significant commonality advantages translating into significant operating costs savings for the operating economics oriented, highly demanding airline customers.

Another game changer technology brought to the market by the Airbus-Boeing duopoly has been the usage of non-metallic structures or composites in airframes. In fact, the Airbus A300; developed in the early 1970s, almost half a century back, was the world's first commercial airliner to pioneer & feature usage of composites in the form of fiber glass reinforced plastics for the leading & trailing edges of the tail fin[2]. Airbus later expanded the usage of composites on the A310 airframe with their application in the engine pylons & other tail sections. The initial A320s had fiber reinforced plastic (FRP) composites constituting almost 10% of the A320 airframe's weight which increased to almost 25% on the A380 and went further up to 52% on the A350XWB.

This usage of composites by Airbus almost 4-5 decades back was enabled & pioneered by Spain based CASA, a noted manufacturer of military transport aircrafts, which had been an Airbus partner. Boeing's first usage of the composites in airframes, on the contrary, can be narrowed down to the 777 built in early 1990s followed by their extensive utilization in the 21st century 787 Dreamliner program amounting to almost 50% of the airframe's weight. However, Boeing did pioneer the utilization & application of fiber glass in commercial aircrafts on the 707 in the 1950s.

The introduction of the world's first twin engine wide body airliners in the early 1970s; the era of quad & trijets; also needs be mentioned as another remarkable achievement for Airbus. The A300 took on the might of the Boeing 747 quadjet, the less fortunate L-1011 TriStar trijet from Lockheed Corporation and the DC-10 trijet. The Airbus decision to conceive, design & build the 'plain jane' A300/310 purely from a utilitarian perspective, pivoted around the concept of operating economics as the core theme, turned to be a big boon and a savior through the tumultuous decade of the 1970s marred by the unprecedented oil crisis. The A300/310 program did leave its mark in the short & medium haul market, despite its limited success in the traditional long haul market, with around 800+ A300/310 aircrafts built & delivered by Airbus with the program later finding flavor in the freighter market as well, as its secondary application.

The A300/310 program also provided Airbus with a ready, wide body aircraft platform to build upon and it ultimately metamorphosed into the A330/340 program later with the A330 retaining the same fuselage section diameter as its predecessor, the A300. However, the Airbus decision to go towards the narrow body side of the market in the early to mid-1980s, with the A320 pursuit, rather than going in the opposite direction to further stretch the A300/310 to develop the A330/340, was one of the key calls taken by the Airbus command in the 1980s given that U.S. airlines were unwilling to take the significant capital risks of buying a wide body aircraft from an unknown European OEM at the time. This made sense given that the list price of the A300/310 in 1984 ranged from $40 to $50 million while the A320 had a list price of around $26 million which simply translated into halving of the overall capital risks for the airlines at the time. Further, the U.S. industry deregulation in 1979 had already set the launch pad in place for the narrow body segment's take off and catapulting of demand growth rates.

Genesis of the 737 Original

Boeing, in the 1980s, had been miffed at the Airbus' decision to launch a FBW enabled scratch up aircraft program to take on the 737-100 & 200; which were almost two decades old at the time chronologically as well as technologically; in addition to the 737 Classic series comprising the -300/-400/ & the -500 variants which had been launched in 1979 & had just entered service in 1984; the very same year in which Airbus had launched its A320 aircraft family.

Boeing, however, had been operating in the reactive strategic mode even at the time of the launch of the original 737-100 & 200 variants despite being a manufacturing giant for commercial jetliners in the 1960s with the 737-100/200 locking horns with the DC-9 from Douglas apart from the BAC-111 & Fokker F28, which were in the final stages of their certification and were expected to be in the air by 1966[3].

To reduce development time while being pushed really hard by the fastidious launch customer Lufthansa; Boeing decided to maintain 60% commonality in structures & systems between the 737 & the 727 with the hallmark of the approach being the usage of the common fuselage cross section. The call taken by the legendary Boeing airplane designer Jack Steiner himself[4] based on some serious pushing & shoving by Lufthansa, who placed initial orders for the 737-100s in 1968, thereby, becoming the first international launch customer for a Boeing commercial aircraft program. It was not surprising given that 737 program, as per Boeing's product strategy, was to supplement the tri-engine 727 in Boeing's short haul aircraft portfolio targeted at short & thin routes.

The original 737, in fact, had been conceived originally as a 50-60 seat aircraft, almost half the size of the 727, with an operating range of up to 1,000 miles based on Boeing's 'detailed' internal market research. However, thanks to the reluctant & scrupulous launch customer Lufthansa, the original plan had to be jettisoned with alterations made to the initial specifications with the seating capacity getting increased to 100 apart from multiple other changes. Another fortunate order from the United Airlines in 1965 led to the first fuselage stretch on the 737 and creation of the ultimately successful 737-200 variant.

The 737-100/200 competed head-on with the DC-8 & DC-9 from the Douglas stable in the narrow body aircraft market with Douglas having launched its single aisle quadjet, the DC-8, powered by Pratt & Whitney's JT3C turbojets way back in the mid-1950s which had entered service in 1959.

The 737 Originals were powered by a pair of Pratt & Whitney's JT8D low bypass turbofans which also powered the Boeing's sole trijet, the 727. The DC-9; which entered service two years earlier than the 737-100 in 1965 powered by the pair of same JT8Ds as the 737; however, maintained a clear market lead over the 737-100 & -200, in terms of order intake & deliveries and had been clobbering them through the initial years. Boeing delivered 334[21] 737-100 & -200 aircrafts over the first 7 years (post entering service), i.e. 1967-1973, while 709[22] DC-9s (more than double) were delivered over 1965-1973 just prior to commercial aviation's heavy battering by the onset of the global oil crisis in October 1973.

It was only after the industry deregulation in 1979 that the 737 program managed to touch three digit delivery numbers for the first time in 1981 after almost a decade-and-a-half of entering service. The commercial aviation market in fact had clicked into top gear in the 1980s propelled by industry deregulation along with softness of the crude oil prices which had been at the rock bottom levels boosting airlines profitability which in turn bolstered aircraft deliveries across OEMs. The aircraft orders & deliveries had been booming for Boeing's 737, which had its second generation Classic series launched in 1979, as well as McDonnell Douglas' DC-9 program, which had its successor launched in 1977, the MD-80.

The 737 effectively was the first aircraft at the time to be certified for Category II approaches capable of precision instrument approach and landing. Developed at a modest budget of $150 million (with which one can't even buy a pair of 737s or A320s today based on list prices) almost over 50 years back under the leadership of then Boeing President Bill Allen, the 737-200 turned out to be almost a pole star for Boeing. In fact, a few 737-200 aircrafts had still been in service with airlines across the less developed & remote parts of the globe till almost 2018 with the operators sold out on the aircraft's remarkable short field landings capability and the ability to land on unpaved landing strips.

This capability of the 737-200 to operate seamlessly from unpaved runways and remote airports was a huge advantage in the aviation world & infrastructure of the 1960s especially outside traditional markets & across emerging markets and was effectively turned into the lynchpin of the 737's sales pitch by Boeing with great special effects aimed at bolstering sales uptake during the 737's initial, tumultuous years after entering service. In fact, only 30 units of the 737-100 were produced ultimately (most of which went to Frankfurt and were operated by the launch customer Lufthansa) with the airlines preferring the larger -200 variant over the 737-100[5].

The original 737s, i.e. the -100 & -200, had engineering & reliability issues as well, as has also been demonstrated by their crash rates, which have been one of the highest[23] across the 737 program generations (with the exception of MAX, of course), which improved dramatically through subsequent generations with the 737NG having the best, overall safety track record on the 737 program.

737 on the 'Chopping Block'

Interestingly, there was also a time when Boeing had considered pulling the plug on the 737 program to somehow cut down on its mounting financial losses. It was during the 1970s, precisely amid the 1973 oil crisis, which had brought many airlines to the brink of bankruptcy. Facing pressures on the order intake front on the 737 program and financial pressures on the other with the company almost on the verge of being broke financially; Boeing had even considered putting the 737 program on the chopping block with contemplation of plans to sell the entire 737 program to the Japanese aerospace industry.

The same was confirmed by Jack Steiner, the 737 airplane designer himself, in an interview given much later to the leading aviation industry magazine, FlightGlobal. "I can't tell you we would have gone through with it, but the intention was there. We were broke" Steiner said[4]. The same has also been ratified in an interview by Peter Morton, who was the marketing manager of the 737 program at Boeing from 1969 to 1974 prior to being an engineer on the 737 and ultimately retiring as a Vice President. He said, "It was a volatile time. The company was in financial straits, and everything was on the table" [20].

This was palatable given that the 737 Original has been the least successful of the four 737 program generations with Boeing having produced only 1,144 737-100/200s over 2 decades of production, from 1967-1988, of the total 10,500+ 737s the company has produced by mid-2020. In fact, only 30 737-100s were produced & delivered to customers owing to deficiencies while the improved 737-200 fared much better and accounted for the remaining 1000+ 737 Original aircrafts produced. The four year period spanning 1970-1973 was the toughest phase for the 737 program with order intake under tremendous pressure while facing stiff competition from DC-9 and Boeing delivering a total of just 111 737s in total over these 4 difficult years with the program's numbers really taking off only post the market deregulation in 1979.

However, fortunately for Boeing, a way out was figured out with the cancellation of the Boeing Supersonic Transport (SST) (which if had gone ahead would have created financial mayhem for Boeing given the onset of the oil crisis in the 1970s) and a ramp down of production rate on the 747 program which created enough elbow room at Boeing for the 737 program or the aviation future (in fact the present) would have been altered to be something else altogether, much in the 'back to the future' trilogy way! In fact, looking at the 737 program's overall production numbers at 10,000+ aircrafts over the course of it's more than half a century long production run, Boeing executives surely would have thanked their stars in retrospect.

Commercial Aviation's Technological Generational Leap in the 1970s:- From Turbojets to Turbofans

The 1970s was a decade of transition & transformation for commercial aviation spearheaded by the convergence of an engine technology revolution with the precipitation of global oil crisis which collectively both from demand and the supply side perspective paved the way for the evolution of much more fuel efficient aircrafts on the aviation horizon.

The decade marked the industry's transition from the fuel guzzling & noisy turbojets to the much more powerful, quieter & fuel efficient turbofan engines starting with the wide body market landscape. The trend was led by the Pratt & Whitney's JT9D high bypass turbofans powering the world's first wide body aircraft program, the Boeing 747, in the late 1960s followed by the GE's CF6 engine, derived from its military counterpart TF39, which gave a further impetus to the change wave in the early 1970s thereby effectively becoming the industry benchmark for the wide body turbofan engines in the process and staying there for decades to come with Rolls Royce's RB211 & PW's JT9D staying firmly in tow.

The technological shift in the narrow body aircraft market in the 1970s, however, was shaped & powered by a market development in the early 1970s which was going to radically alter the dynamics of the medium thrust aircraft propulsion market forever. It was the creation of CFM International; a 50:50 owned joint venture between General Electric and SNECMA; which has come to rule the narrow body aircraft propulsion market for almost half a century now.

The CFM International's iconic CFM56 engine program ultimately became the narrow body segment's gold standard in performance & reliability after a horrendous market debut marked by an almost half a decade- long orders famine in the 1970s spanning 1974-1979. The jinx however was broken & followed by a soaring launch ride provided by the USAF's KC-135 program via the re-engining route. The CFM56 drove the re-engining of the 1950s era, DC-8 quadjets; powered by the Pratt &

Whitney's JT3D turbojets (later models); for a number of airliners while also catalyzing the genesis of the Boeing 737's second generation, the 737 Classic in the late 1970s.

The generationally advanced, clean sheet Airbus A320, too, was powered by the CFM56-5 series engines in the mid-1980s. The CFM56 engine also drove the re-engining of a number of in-service narrow body aircraft programs powered by the Pratt & Whitney's low bypass JT8D turbofan engines in the 1980s, which included, the 737-200, 727-200, DC-9 and the MD-80.

The International Aero Engines (IAE) consortium, comprising leading engine manufacturers[1]; led by Pratt & Whitney, Rolls Royce, MTU & the Japanese Aero Engine Corporation; launched its V2500 high bypass turbofan engine in the 1980s looking to grab a share of the narrow body growth pie as a rival to the CFM56 & to check its virtual market dominance, with the V2500 receiving FAA certification in 1988, and thereby becoming the arch-rival of CFM-56 in the medium thrust propulsion market.

737 Classic series was the second generation of the 737 program comprising the larger 737-300/-400/-500 variants and was launched by Boeing in 1979 aimed at capitalizing on the upcoming growth wave in form of the industry deregulation in the U.S. market in 1979. Sporting the latest high-bypass CFM56-3B-1 turbofan engines while also incorporating latest avionics upgrades as well as changes to the wings; the 737 Classic series featured improved operating economics with a lower fuel burn rate, reduced noise levels and increased range as well as passenger capacity. This was done by Boeing to take on the rival MD-80 (launched in 1977)

---

[1] IAE also had FiatAvio as a program partner on the V2500 originally, as reflected by the V in V2500 which stood for the five partnering companies, but it exited in the initial stages itself as partner but remained on-board as supplier.

effectively.

The aviation landscape had also changed drastically by this time with the emergence of Airbus which had effectively made its market debut with the world's first twin-engine commercial wide body aircraft, the A300, which entered service with Air France in 1974 followed by the unveiling of its shorter variant, the A310, launched in 1978, which too was in the medium wide body aircraft market segment.

At this stage, it is important to highlight a peculiar 737 issue which although was MacGyvered successfully at the time of the development of the 737 Classic series but was going to lead to further issues & challenges at the time of the 737's generational revision in the 21st century. The wings of the 737 were low and much closer to the ground which had necessitated structural modifications to the CFM56-3 engines selected by Boeing to power the Classic series. CFM International & Boeing sorted out the issue by changing the shape of the engine slightly by reducing the fan size leading to reduction of the overall bypass ratio & the thrust output and mounting the engines ahead of the wing rather than under it while moving engine accessories to the side of the engine pod (more on it later).

737 Classic – A Decent Success

Entering service in 1984, the 737-300 received a grand reception from the airlines customers and the aircraft ultimately turned out to be a decent success for Boeing rising further in popularity by the turn of the century. The 737 Classic series did really well than its predecessor, in terms of sales tally for Boeing, which produced & delivered around 1,988 737 Classics over the 1984 to 1999 period of which 1,113 aircrafts were the 737-300s with hardly 486 & 389 units of the -400 & the shorter -500 variants getting ultimately built[5]. The shorter 737-500 variant was in turn created based on customer demand as a modern replacement of the original 737-200.

The A320's Ascent as a 'Potent Force' in the Narrow Body Segment

The Airbus-Boeing rivalry in the narrow body market segment had begun by the mid-1980s with the launch of the A320 and reached its zenith in the 1990s marked by multiple hard fought aircraft deals with airlines involving significant haggling over pricing & discounting. In the battle for narrow body market supremacy, the A320 built its sales pitch around its significantly lower fuel burn rate & better operating economics based on its CFM56/V2500 engines in a head-to-head comparison with the 737-300 as well as the MD-80.

Airbus also claimed that with a long & thin wing, which also had a better aspect ratio, the A320 offered better aerodynamic efficiency than the competing 737 as well as the MD-80. Boeing had initially ignored the emergence of Airbus and allowed it to effectively create the beachhead it had been looking for desperately. However, Boeing reacted strongly to the European invasion on its terrain later in the mid-1990s and tried to contain Airbus by mounting an unsuccessful counterattack as well. Boeing's reactivity thus effectively laid the foundations for the establishment of a market duopoly in commercial aviation.

The creation of a defanged Douglas post its merger with McDonnell in 1967, which mostly saw variants of the original DC-9 developed into the MD-80, MD-90 and MD-95[2] over the next two decades under a derivative product strategy with limited market success and not even a single clean sheet program, effectively put Douglas on the runway to extinction while also weakening the market competition for Boeing. The exit of Lockheed

---

[2] , marketed as the Boeing 717 post acquisition of McDonnell-Douglas by Boeing in 1997

Corporation from commercial aviation, post the tumultuous decade of the 1970s, with the L-1011 TriStar turning out to be its last commercial jetliner, made Boeing an even more dominant force in commercial aviation.

Post the acquisition of McDonnell Douglas ultimately by Boeing in 1997; spurred by the ensuing defense market slump in the post-cold war world, with defense spending more or less having evaporated completely; it became a virtually all out Boeing Vs. Airbus aerial battle in commercial aviation.

The McDonnell Douglas acquisition, however, did add tremendous value to Boeing on the defense side of its business. The deal brought in some of the most venerable & top of the line military aircraft programs which are still in service even to this day, especially, the F-15E Strike Eagle & F/A-18 Super Hornet combat jets, the C-17 Globe Master III strategic airlifter and the AH-64 Apache attack helicopter. Additionally, through the acquisition Boeing easily & effectively eliminated a weaker but formidable rival in the commercial aircraft market.

A multi-decennial, head-to-head comparison between Boeing and Douglas/McDonnell Douglas on single aisle aircraft deliveries shows just that effectively with Boeing having delivered a total of around 3,577[24] narrow body aircrafts belonging to the 737 Original, Classic & NG generations over the four decades spanning 1959 to 1999 while Douglas & later McDonnell Douglas delivered a total of 2,834[24] narrow body aircrafts (comprising DC-8, DC-9, MD-80 & MD-90 and their variants) over the same period. Apart from 737s, Boeing also delivered almost 1,000[24] units of 707/720 quadjets and around 1,832[24] 727 trijets.

Boeing & McDonnell Douglas, thus, collectively delivered around 9,000+ narrow body aircrafts over these 4 decades with McDonnell Douglas accounting for almost a third of the total aircraft deliveries. Taking the Airbus A320 family into the equation, the total narrow body aircraft deliveries by the three OEMs crosses the 10,000 units mark effectively with Airbus delivering 1,142[25] aircrafts belonging to the A320 family over 1988-1999 with 792[25] of them being the A320.

The technologically superior A320 had been gaining roots in the commercial aviation market in the 1980s following the initial launch ride provided by the European carriers and piggybacking on them, it ultimately succeeded in on-boarding many airline customers & operators on-board beyond Europe, especially in the U.S. market, leveraging its significant fuel efficiency advantages, emanating from the CFM56-5 series engines, over the in-service fleet of 727-200 trijets, DC-9s and MD-80s powered by Pratt & Whitney's low bypass JT8D engines. Add to that the larger seating capacity & way longer operating range and the A320 clearly had an overmatch, even over the 737-400. Airbus had been desperately trying to make inroads into the U.S. market since the launch of the A300/310 wide body aircrafts in the early 1970s without much success.

Airbus, in fact, had been maintaining an inventory of almost two dozen completely built but unused A300 aircrafts at its Toulouse plant in September 1984 amid softening of the overall market demand. The only major breakthrough Airbus had in the U.S. market so far (since the launch of A300/310) was the deal signed with Eastern Airlines in 1982 for procurement of 34 A300 aircrafts.

Besides that deal, Airbus' sales track record in the U.S. market had been dismal with the company having sold only 4 A300s to Northeastern Airlines & a single A300 to Capitol Air. However, some of the most crucial deals for Airbus behind enemy lines were yet to come & were on their way. One of them was the on-boarding of Pan American World Airways (Pan Am) by Airbus in 1984 with a $1 billion deal for procurement of 28 wide as well as narrow body aircrafts, including 16 A320s along with options for another 34, under purchase or leasing agreements (with the aircrafts to be delivered over the 1987-1990 period) with the deal providing Airbus the much needed beachhead in the strategic U.S. market. However, the deal that provided the much needed critical mass to the chain reaction underway at Airbus sales reactors based in North America, run by a sales maverick named John Leahy, was the Northwest deal signed in 1986 for 100 A320s worth $3.2 billion[7]. The deal; a leaf out of Leahy's highly unconventional, exhaustive and effective sales playbook; ensured that the A320 was going to inevitably be in the lead cast in the U.S. & global aviation market's one of the biggest blockbusters in the making! More on Maverick and his playbook & exploits a little further ahead!

The Pan Am beachhead deal was covered superbly by The New York Times (NYT) in an article published in the September 14, 1984 edition with the headline, "$1 Billion order by Pan Am is a lift for Airbus[6]", underscoring the strategic significance of the deal given that Airbus had been struggling to penetrate the U.S. commercial aviation market, the largest in the world and accounting for almost 50% of the global market at the time, for over a decade.

The NYT article had further pejoratives in store for the deal, capturing & echoing the general market reaction it evoked, by describing it as nothing short of a 'coup'. It further read, "The sale to Pan Am is a coup for Airbus - a consortium of French, British, West German and Spanish aircraft makers that has struggled for a decade to penetrate the American aircraft market. The agreement also represents a defeat for the Boeing Company, which had fought Airbus hard for the Pan Am order[6]". Airbus North America President & CEO at the time, Patrick Croze, delivered the victory speech, "the Pan Am order is a very clear indication that Airbus made a sound strategic decision in building a narrow-body passenger jet as a means to attract American carriers[6]". He further said: "The pact indicated the convergence of Pan Am's present fleet needs with what Airbus could offer. Airbus absolutely had to add a narrow-body jet to its line if it were to be competitive in the United States market, while Boeing felt that its current products could carry it through this period[6]." referring to the industry deregulation act which had been passed in the 1978. It was the first initial wins for the A320-200 competing against the 737-300 which had been offered by Boeing to Pan Am along with the 767.

However, Pan Am had been booking losses for 3 consecutive years prior to the year 1984 with the carrier's cumulative tally of losses pegged at almost over half a billion dollars amid difficult market conditions forcing it to operate in the cost cutting mode. Thus, price discounting by Airbus could also have swayed the Pan Am's decision to buy aircrafts from the European OEM, a real possibility which had also been speculated by multiple industry analysts at the time. Coming to the analysts take at the time, Robert Kugel, a Morgan Stanley & Company analyst described the deal as a catapult for Airbus, "It's an important win for Airbus. It says conclusively that they will be a serious competitor in the future along with Boeing and McDonnell Douglas[6]".

Airbus had, in fact, as part of the terms of the agreement, offered to finance the acquisition of airplanes by Pan Am as well with the European OEM going the extra mile for its first American airline customer offering the additional stretch of runway & fire tenders in case of a potential hard landing or crash which ultimately, however, did not prove enough!

The significance of the Pan Am deal for Airbus could not be overstated while it also gave the European airplane maker the much needed breakthrough in taking on Boeing's might in the U.S. market right at the outset of the A320 program launch. For Boeing, it was nothing short of an invasion as Pan Am had been one of its top customers having ordered over 200+ aircrafts since the late 1950s. Pan Am had also been amongst the launch customers for the iconic 747 and had been operating the largest fleet of the 747s maintained by any U.S. carrier at the time with the Pan Am's 747 fleet strength standing at around four dozen 747s back then accounting for almost three fourth of the Pan Am's total seat miles capacity at the time.

Pan Am had planned a 3 tiered approach to its fleet structure with the 747s giving top cover at the upper end of the spectrum with their usage reserved for long range flights followed by the Airbus A310-300s positioned right in the middle with intercontinental range flights as their chosen application. The Airbus A320s were planned to cover the lower end of the spectrum with these new economical machines to be deployed to undertake short to medium range flights mostly covering U.S. domestic routes.

Pan Am, one of the most profitable airlines of its time, however, had sunk from financial distress before it could even take delivery of the first A320s it had ordered. Pan Am, however, did not go down in history without having played a small little cameo of its own and creating a little piece of history of its own. Pan Am has to its credits the distinction of being the launch carrier for the Boeing 747, the queen of the skies, and also the U.S. launch customer for the Airbus A320, the almost queen of sorts of the narrow bodies with its rapid ascent to the throne. Both of these have been iconic aircraft programs which are still very much there on the horizon despite the cyclical swings in market vicissitudes!

A320's Disastrous Take-off to its Flying Career…

The Airbus A320 actually had a terrible start to its flying career in 1988, right after its entry into service. The A320 program had just entered service with aircraft deliveries to customers commencing in April 1988 starting with launch customer Air France and Airbus had planned a demonstration flight of the A320-111, the world's first civilian fly by wire aircraft, at the Habsheim Airshow, which was to be held in June 1988. The demonstration, which was going to be the A320's first passenger flight, was to take place in front of thousands of spectators with the chartered flight, Air France Flight AF296, being amongst the first three A320 aircrafts delivered to Air France. The demonstration flight was also to be boarded by journalists apart from some promotional contest winners.

The demo flight, with 136 people on board, took off on June 26, 1988 from the Charles de Gaulle Airport and as per the plan[8] it was to make a low-speed flyover, as part of the demonstration, followed by a stopover at the Habsheim aerodrome on its way to the Basel-Mulhouse Airport which was to be its final destination. The A320-111; captained by a seasoned Air France technical pilot with over 10,000+ flight hours[8] under his belt; was originally scheduled to perform the low-speed flyover with landing gear down at an altitude of 100 feet but it actually ended up doing so at the 30 feet level. The brand new aircraft, thus, brushed the tree tops at the airfield's end and ultimately crashed right in front of a huge en masse of media, spectators and the shutterbugs shocking the aviation world.

Fortunately, 133 of the 136 people aboard the aircraft survived. However, 3 of the passengers succumbed to asphyxiation arising from smoke inhalation and around 50 were injured. The post crash investigations, led by the French Air Accident Investigation Bureau, pointed in its report[8] towards human error attributable chiefly to low altitude, slow speed, engines speed at flight idle and late application of go around power. However, the report added more fuel to fire rather than dousing it and created more controversy with the flight deck crew involved in the crash questioning the responsiveness of the aircraft's fly by wire controls apart from allegations made by the French pilots union, SNPL, and media factions alleging the A320's role as a contributory cause[8, 9] to the accident which were in turn defended vehemently by Airbus in its official response to the same.

The potential ramifications of the accident & stakes involved for Airbus post crash, with reference to the market prospects for the A320, were huge given that Airbus by mid-1988 already had over 500 aircraft orders for the A320 program with delivery slots completely booked through 1993[10]. This was simply one of the worst possible starts for a radically different & generationally advanced aircraft program boasting of ground shaking technologies coming from an industry challenger!

A Maverick with the Vision, On a Mission and the $1+ Trillion Difference it ultimately made

Another key move from Airbus, which turned out to be a real game changer, was made in January 1985 with the appointment of John Leahy to its North American sales team, which he soon was going to lead. John Leahy had been a former licensed commercial pilot as well as a flight instructor and had been working with Piper Aircraft heading the marketing department. He was bound to move to Europe having just accepted a position from Piper as the Director of Sales for Eastern Hemisphere to be based at Geneva, Switzerland.

Fate, however, did have Europe deployment plans for him but they were to materialize exactly a decade later and precisely just prior to moving out of the U.S. he got a call from the recruiter appointed by the Airbus and for a change Leahy got convinced by the head hunter to work with a truly

European company, Airbus as recounted vividly by Leahy in an interview post his retirement[11].

Leahy recalled precisely as to what the head hunter had told him; in a conversation with Leeham News' Scott Hamilton, who got the maverick to share some of his old trade secrets & adventures; while reflecting upon his career at retirement. Leahy recollected[11] the recruiter having told him in his aggressive pitch that working in Europe for an American company would be tantamount to a European deployment on an American military base for the U.S. military personnel and exhorted him to instead switch to a true blue European player for a truly European work experience.

The pitch worked on Leahy albeit followed by the paradox that he had to continue working in the U.S. despite being hired by an aboriginal European company. If Boeing had the crystal ball at the time, it would surely have pressed for leveling of mutiny charges against the head hunter and would have surely got the deal termed against 'national interest' and would have subsequently come up with something on the grounds of raison d'état to somehow stop things in their tracks & thwart them in order to prevent the colossal damages these set of events were going to lead to for Boeing over the next few decades!

The first key sales deal Leahy cracked after boarding the Airbus deck was signed with Northwest Airlines, which had primarily been a Boeing operator, for the A320s in 1986. Therein, he used the "Buy small, think big" strategy, as outlined by him in an interview with the Seattle Times[12]. Only a few of the airlines in the U.S. market by then had placed orders for the A320, including, Eastern, Continental & a financially turbulent Pan Am and the airplane itself was still under development. He told Northwest that they could place a firm order for 10 A320s while getting a bulk discount equivalent of a 100 airplane order. But "we'll give you delivery dates for

100[12]." If they didn't like the Airbus planes, he told them, "That's it. You're stuck with 10. We'll take the risk on the rest of them." "Sure enough, they loved the airplane," he said[12]. "They didn't take just 100. I think they got up to 145."[12]

The deal for 100 aircrafts was worth a total of almost $3.2 billion with deliveries scheduled from 1990 through 1995; however, only 10 of the aircraft orders were firm. Northwest had just merged with Republic Airlines in October 1986, which was one of the largest industry mergers by then, aimed at countering United's acquisition of Pan Am's Pacific operations a year back in 1985. Northwest, thus, had been in the process of streamlining its operations post merger & wanted flexibility and the deal was structured by Airbus in a manner to offer just that.

The A320 deal included firm orders for the first 10 aircrafts followed by six blocks of 15 aircrafts each accounting for the remaining 90 aircrafts. As per the deal, Northwest, post the delivery of the initial 10 aircrafts, could either confirm or cancel all or any of the remaining aircrafts on order in a block-wise manner with decisions to go-ahead & move towards confirmed orders required for each block of aircrafts separately. The confirmation for the first block, thus, was required in 1987 for deliveries scheduled for the early 1990s. However, all of the order blocks had been confirmed by Northwest by 1990.

This is how he started his outlandish sales campaign which was followed by signing of a similar aircraft deal with United and he took Airbus ultimately from nowhere to becoming Boeing's arch rival & nemesis, especially in the narrow body aircraft market. By 1988 Leahy had become the President of Airbus North America entrusted with the responsibility of making deeper inroads into the U.S. commercial aviation market which had hitherto been dominated by Boeing and he surely delivered by zeroing on to & converting multiple airline customers, including even those who were staunch Boeing loyalists, to Airbus camp.

Leahy, thus effectively torpedoed Boeing's business deals tactically by often undercutting them on pricing, sweetening deals further with financing and trials along with no costs and no questions asked returns. Powered by a series of such aggressive, daring & disruptive market exploits; Airbus booked cumulative orders for 825[25] aircrafts (including A300/310 wide-bodies) worth $61 billion collectively for 1989 &1990, of which 50% or 446[25] aircraft orders were for the A320 family, and Airbus, thus, took-off to the skies!

Airbus' board & the high command, too, surely deserve praise & due credit in having empowered their maverick general, rather than stifling him, to unleash hell on a heavily fortified enemy stronghold after successfully invading it like a skilled Special Forces Operator maneuvering his way in while operating right below the enemy radar. He conquered & Airbus in turn gave him a global play.

In 1994, Leahy was appointed the Chief Commercial Officer and along came his deployment orders to Toulouse, France wherein he was to play the lead role in taking Airbus from a relatively minor player in the market to a worthy competitor & Boeing's almost equal by the end of the twentieth century. Let's review that tipping point as described by the maestro himself later in another of his farewell souvenirs given again to Leeham News[13].

In January 1995, less than a year of moving to the headquarters, Leahy had presented his ambitious goal to Airbus board of catapulting from a low double digit market share in the mid-1990s to reaching a 50% market share by the year 2000, in less than 5 years, for which he was even scoffed at initially by the board members who later acceded reluctantly. Leahy gave a verbatim account of the conversation he had back then with the board and it's hugely significant from the grand Airbus vs. Boeing strategic context going forward from here on. They said "that sounds very motivating for your sales team. But we want to know what is the real and realistic goal of market share that we should be aiming for?" [13] "I paused, and I said '50%. That's what we need to do.' They started laughing. Somebody said, 'you can't be serious. Twenty-five percent, maybe 30% would be a more reasonable, more attainable target[13]." Eventually, 50% was locked on as the new target to be chased nevertheless, much to the skepticism of the Airbus executive board.

Leahy exhorted the Airbus' board to not make the same mistake and built his 50% target argument around McDonnell Douglas' strategy for its commercial business spearheaded by the adoption of a sporadic & piecemeal approach underscored by the pursuit of smaller, easy hunts in a limited part of the ocean which in turn set itself up perfectly on the road to eventual extinction. Leahy, thus, wanted Airbus to not fall prey to that kind of a self-destructive mindset & strategic posture.

However, in the first year Airbus managed a mere 18% market share and Leahy had been expecting fire & fury from the board in the next board meeting. The Airbus board meeting, which took place exactly a year later in January 1996, however, turned out to be a much different affair, as per Leahy[13], as the board was quite comfortable as it had in fact treated it as a long term target and Leahy was ultimately spared from the heavy shelling he had actually anticipated. Airbus, in fact, effectively gave him a free hand and he made the most of it while being in command with a series of ensuing victories in multiple, high stakes aerial dogfights leading to the acquisition of a number of strategic aerial targets eventually!

By the year 1999, with the end of a strategic & game changer decade of 1990s for the European consortium, Airbus had reached the 50% milestone with a year to spare, a target set by Leahy which had seemed unthinkable to most of the Airbus executives four years back in 1995. The Maverick, with his unconventional playbook of high G maneuvers, thus had pushed the much more orthodox Airbus board to go for what they thought would be a wild 'goose' chase but he proved it otherwise and ultimately saw them through, coming back triumphantly, in turn, to buzz the tower!

By the time Leahy finished his long & illustrious career with Airbus in early 2018, he had "overseen sales of almost 15,500 Airbus Airplanes worth around $1.7 trillion"[14] to airlines worldwide, a huge & mind boggling number by any means of imagination!

The Defining Moment and the 737NG

The deal that became the defining moment in the Airbus-Boeing arch-rivalry was the Airbus sales deal signed with United Airlines. By the early 1990s, Airbus had effectively gobbled up McDonnell Douglas' market share and had clinched the second spot in aircraft deliveries with the struggling & cash strapped veteran U.S. aircraft manufacturer finding it tough to invest towards new product development stifled further by its piecemeal approach to strategy.

The game changer moment in the Airbus-Boeing rivalry, thus, came when Boeing's biggest & top airlines customer, United Airlines, struck an unexpected $2.4 billion deal[19] with Airbus in July 1992. As per the agreement, Airbus was to lease 50 A320s (including options for another 50) to United Airlines at $28 million per aircraft, which was almost $4 million less than even the production cost of the A320 at the time, while the Boeing 737-400 had a prevailing list price of $29 million[15] along with its relatively much limited range as compared to the A320.·

Boeing was enraged at this as it firmly believed that the 737-400 was far superior to the A320 despite its obvious operating range limitations against the A320 and that United would never buy anything but the 737. This deal had shaken Boeing to the core and a shell-shocked Boeing launched its counterattack in November 1993 announcing its plan to develop the third generation of the 737 program (which was almost two and a half decades old by then), designated the 737 Next Generation or NG in short, which would overcome the range & capacity limitations of the older 737 Classic series and would be able to compete more effectively with the A320. "We were probably responsible for the [Boeing] 737NG because a lot of people thought the 737 Classic could outsell the A320,"[13] Leahy commented on the deal with United. "At least the guys in Seattle thought that. But when the dyed-in-the-wool, 100% Boeing customer United switched over to the A320, that's when they panicked and decided they needed to update the 737[13]"

The 737NG series comprised of the 737-600/-700/-800/-900 aircrafts with seating capacity ranging from 108 to 215 passengers. The 737NG aircrafts were comprehensively revamped and featured a larger, redesigned wing with a wider wingspan, incorporation of glass cockpit and redesigned cabin layouts & configurations. In terms of performance specs, the 737NGs had longer operating range & increased fuel capacity along with a higher MTOW, courtesy the new pair of CFM56-7 series engines, which were to power them. However, the new 737NG was still devoid of the fly by wire capability which still kept the generational gap between the 737 & the A320 aircraft families somehow very much intact!

"In Warfare or Business, take a long-term view while planning strategy and a short term one while executing it" – Anonymous

THE 'LOST WINDOW OF OPPORTUNITY' & SOME 'TOUGH LESSONS'

"In the midst of chaos, there is also opportunity" – Sun Tzu, The Art of War

In a way, Airbus, with the United deal, had forced Boeing's hand into taking the call for the genesis of 737's third generation and this was the very moment in time which was to repeat itself in exactly the same manner almost 2 decades later when Boeing again chose to go the re-engining way to create the 4th generation of the 737, designated MAX, in 2011 but with another operator, American lurching in the background.

However, this time round in 1992, the decision to re-engine the 737 to create 737NG, rather than taking a clean sheet approach for 737's succession, was undertaken given the need to take on the A320 in a timely manner which clearly favored the re-engine route. Secondly, Boeing at the time had been working on the development of its ambitious 777 program which was going to be the largest wide body twinjet ever built and had its maiden flight ultimately in 1994.

Boeing, thus, had its engineering & financial resources already invested fully in the high priority program with the 777 development ultimately costing Boeing much more than the originally budgeted $5 billion to develop. Lastly, the commercial aviation market in the early 1990s was in a marked downswing with the North American market operating at almost 40% of the usual capacity.

Boeing, in fact, had been studying the prospects of launching a new, clean sheet aircraft program extensively in the mid-1980s, designated as 7-7 and later 7J7, targeted at the short & medium haul segment, with a 150-seat sweet spot at the core, right when Airbus had launched the A320. Additionally, Boeing's 757 had just entered service in 1983 and had received a weak market reception from the airlines which had instead been leaning towards smaller aircrafts in a deregulated U.S. market environment of the 1980s.

Termed 7J7, the program had been elaborately conceived for domestic operations and was to be powered by an exotic propfan propulsion technology which alone promised delivering almost over 40% improvements in fuel consumption as compared to conventional turbofan engines of the generation with similar specifications. However, pitching fuel efficiency as a key proposition was quite incongruous in a low crude oil price environment of the 1980s which had seen oil prices virtually plummeting after having just skyrocketed during the oil crisis of the 1970s.

The 1980s had, in fact, been a chaotic decade for the industry with oil prices abruptly changing polarity and plummeting led by an OPEC supply glut which in turn improved airlines profitability. Further, the deregulation of the airlines industry in the U.S. in 1979 opened the market floodgates to new & spirited flying cowboys which brought in a significant increase in competitive intensity in the market.

To capitalize on the same, Airbus rode in with its original A320 narrow body in the mid-1980s packing a hefty punch with its next generation technologies suite. The centre of gravity of Boeing's product strategy focus & game plan back in the early 1980s had been the middle of the market with its latest pair of clean sheet fraternal aircraft twins, the 757 & 767, having just entered service in the early 1980s to take on the first pair of hardy, rugged & spirited French Camargue workhorses, the Airbus A300 & A310 respectively.

The 7J7 (the 7-7 originally) program had been conceived by Boeing in 1983 aimed originally as a replacement for the ageing 727 trijet, which was almost on the verge of being retired, which would have created a gap in Boeing's product portfolio between the single aisle 737 & the 757 positioned right in the middle of the market. The point was pertinent given that the 737-100, the smallest aircraft of the 737 Original series, had a seating capacity of 118 passengers while the larger 757-200 had really long legs & could easily carry over 200 passengers across the Atlantic.

Further, in a deregulated market environment in the U.S. the smaller aircrafts were gaining ground over larger ones with the economics favoring them overall which resulted in the weak market debut of the 757 which had just entered service in early 1983. The 757 was followed by the first member of the 737 Classic series, the 737-300's EIS, in late 1984.

The 150 seat 7J7, thus, hypothetically would have filled the emerging void in the portfolio perfectly. Additionally, Scandinavian Airlines (SAS) was looking to replace its fleet of DC-9s at the time and prodded Boeing for a potential aircraft concept for replacement and conveyed its eagerness to be the launch customer which gave further impetus to Boeing's pursuit of the 7J7 program.

Boeing had proposed significant utilization of the Aluminum-Lithium alloys and carbon fiber based composites on the 7J7 airframe to reduce weight while also planning to incorporate a glass cockpit fitted with LCD panels and fly-by-wire flight controls[8]. In terms of cabin design, Boeing had outlined the plan to use a twin-aisle based seating configuration to reduce enplaning & deplaning time as the primary plan which would have come at the expense of an increase in overall aircraft weight while keeping the backup plan intact for a single-aisle based fuselage and cabin configuration and Boeing practically kept vacillating between them. Additionally, Boeing was also looking at creating two aircraft variants with different gross weights.

Theoretically, at the outset, Boeing was looking at the 7J7 as a 727 replacement with plans for a similar seating capacity of around 150, a similar operating range of around 2250+ nmi[7,8] (with the 7J7 base variant), a similar MTOW of around 160,000 lb -170,000 lb[7,8] but with a different six or seven-abreast seating layout in a twin aisle configuration. However, with the higher gross weight variant, Boeing was almost heading towards the middle of the market, targeting the operating range of the 757 at almost 4000+ nmi[7,8] with a six-abreast seating configuration. The fuselage width on the 7J7 was also being planned to be almost 15% to 30% more[7,8] than the 727.

The 7J7 concept, in terms of specifications unveiled in August 1987, before the program was scuttled eventually, had a close match in Airbus' portfolio as well and it was the A320 launched just a couple of years back in 1984 with a similar MTOW of 172,000 lb, a typical 2-class seating of 150 passengers and an operating range of 3,300 nmi[2]. However, dimensionally the 7J7 concept was positioned somewhere beyond the A320 & closer to its larger sibling, the A321, which was non-existent at the time. However, the 7J7's wing span & wing area were dimensionally even larger than the A321, based on the August 1987 specifications.

Boeing evidently was looking to outgun the A320 on technological superiority front spearheaded by the planned incorporation of the way more fuel efficient propfans. Robert R. Waggener, Boeing's customer manager at the time, gushed in an interview given in 1986 about propfan technology, "The technology that will be ready by 1989 (referring to the A320) will make an aircraft 9% cheaper to run than today's airliners. By 1992, the improvement could be up to 60% (referring to propfans powered 7J7)."[10] A propfan powered 7J7 incorporating cutting edge technologies, thus, theoretically could have ruined the business case for the A320 had the propfan technology commercialized & delivered on the projected fuel efficiency gains, noise levels & overall reliability and it could, in turn, have been a befitting reply from Boeing to the A320. However, the decision was subject to market forces' jurisdiction with the ball effectively being in Boeing's court.

Boeing, with the 7J7, thus, had been proposing an irresistible value proposition to the airlines theoretically with the propfans' huge fuel efficiency advantages over traditional turbofan engines of the time with the same eliciting keen interest in the 7J7 from across the Atlantic in Europe to markets as far as Japan.

The Japanese, in fact, proposed to even be a strategic investor in the program by signing a letter of understanding to own a 25% stake[10] in the program in 1984 worth $2 billion and offered to even fund the development as well if Boeing powered the 7J7 with the IAE's V2500 engine program which had just been launched and had a major involvement of the Japanese Aero Engine Corporation as a key program partner[1].

The Japanese were looking for a 1988 entry into service of the "YXX" program, as they had termed it. Boeing, however, had other plans in the offing as the company was looking for a technological leap generationally and had been highly enamored by the propfan technology along with its enormous potential of double digit fuel savings.

Boeing ultimately ruled in favor of propfans over conventional turbofans by going for & choosing the GE's unproven, gearless Unducted Fan (UDF) propulsion system for the 7J7, much to Japan's chagrin. The move; which further delayed the proposed entry into service of the 7J7 to 1992; simultaneously also infused significant, unnecessary risks into the program as well, given the additional time horizon required for propfan technology's maturation & certification.

Boeing showcased the 7J7 concept at the Paris Air Show in 1985 and announced that the company would start booking orders for the 7J7 from 1987-88 onwards with the proposed entry into service of the 7J7 slated for the first half of 1992[2]. The resilient Japanese, however, were determined to make deeper inroads into commercial aviation and hopped back on to the 7J7 program with Boeing formally announcing their 25% stake in 1986[10] worth $2 billion[10].

There were, however, some other passengers as well apart from the Japanese, who had boarded the 7J7 flight, which included, the British in form of Shorts, Swedes in form of Saab-Scania along with the Australians in the form of Hawker de Havilland, with almost all of them having chosen the economy class with their investments staying limited to small, single digit percentages. However, this inadvertently had turned the 7J7 development formula & approach on its head into almost an international collaborative effort.

Boeing, however, could not finalize a final specification for the 7J7 with the company initially announcing a two version development plan for the 7J7 comprising a 150-seat larger version which was to be joined by its smaller 110-115 seat sibling powered by the propfans. The larger, 150-seat variant was to be powered by IAE's proposed V2500 SF SuperFan, a geared turbofan engine, which was to be a variant of the V2500 with a proposed 20% better fuel burn rate underpinned by a much higher bypass ratio of almost 18:1 to 20:1. The SuperFan was going to be a conventional, under-wing mounted turbofan engine, a program which was eventually terminated by the IAE's board eventually.

Boeing, however, a year later in 1987, jettisoned the development of the smaller 7J7 variant altogether and instead chose to go for a 737 derivative to address that segment while also going back to UDF propulsion for the 7J7. Additionally, the two European carriers and United Airlines, who had been the most avid prospective customers of the 7J7, were on different pages & had absolutely contrasting requirements. The European airliners, SAS & British Airways, wanted a smaller 140-seat airplane with SAS looking to order almost 100 planes while United Airlines wanted a stretched 170-seat variant of the 7J7 with a potential 100 aircraft order of its own in the pipeline.

Boeing, however, was faced with a serious dilemma and could not take an outright call on it given that developing the 7J7 as the larger 170 seat variant at the outset could have diminished the program's future potential with serious doubts hanging over the unproven UDF engine's ability to be able to power further stretched, larger & heavier derivatives of the 7J7. Further, the 7J7, which had started its circuitous journey originally as a potential 727 replacement for Boeing, was later, by the start of the 1990s, was also being considered as a plank for a potential 737 replacement with a major strategic recalibration of the program, which too was ultimately squandered.

This dilly dallying by Boeing with frequent program revisions, design changes and doubts over the unproven propfan technology's commercialization along with its potential operating economics as well as noise levels dissuaded the airline customers from committing to the program in a market environment characterized typically by oversupply which, in turn, further increased Boeing's lack of surety pertaining to taking the program further into production stage. Gordon McKenzie, Manager for new technology at United Airlines aptly captured the market's reaction to the 7J7, "One week it was a single-aisle 90-passenger airplane, the next a 180-seat twin-aisle design. We saw things as being very fluid."[3]

Boeing's marketing road show for the 7J7 program in 1987, in fact, provided the much needed reality check to the company with the same getting a lukewarm response from airlines as Boeing had practically sidestepped them and had been unilaterally defining aircraft specifications rather than taking them onboard at the design stage itself.

This was a crucial lesson learnt by Boeing the hard way with the eventual floundering of the 7J7. Ultimately, the 7J7 program, in which Boeing had invested significant resources towards development, apparently turned out to be more or less an internal developmental exercise with Boeing eventually setting it on a slow death course with gradual reduction in funding & resource allocation rather than pressing the hard kill button outright. Following this, Boeing recalibrated its strategic compass & focused on to the 777 program in 1989 as part of a major & sweeping product strategy shift.

Boeing and its engineering team had conceived the 7J7 as a revolutionary, next generation airplane and had embellished it with all the possible bells & whistles but had missed out on a core ingredient, the airline customers, which were going to actually use and run their businesses on it. These disgruntled customers, on the contrary, had a very different take on the 7J7 as they just didn't appreciate & approve of the Table d'hôte set up being hosted by Boeing.

An article appearing in the May 1987 edition[11] of the Flying magazine does mention that the airlines had told Boeing in 1983 itself that the 7-7 (later designated 7J7) was too expensive and had told Boeing that they would only be willing to pay $28 million[11] per aircraft, which was in turn going to be a long shot, given the broad range of next gen, exotic technologies Boeing was looking to pack into it. At $28 million, there was still going to be almost $2 million difference between the 7-7 and the A320, which was priced (list price) at $26 million in the mid-1980s, which Boeing was hoping to bridge with the much higher fuel efficiency of the 7J7 coupled with the added advantages of the twin-aisle configuration, in terms of faster enplaning & deplaning times, which would have added further to the bottom lines of the airlines.

The article further mentions that Boeing had started working towards designing & developing the 7J7 around the given $28 million price ceiling etched in stone. Boeing's estimates had pegged the propfan powered 7J7 to be almost 50% more fuel efficient[11] than the existing 727 trijet it was going to replace.

Boeing had appointed Alan Mulally; who later went on to fame becoming the Ford Motor Company's CEO and turned it around successfully; as the program's Director of Engineering. Alan said later, quoting Boeing chairman Thornton "T." Wilson that "the 7J7 was one of the best research and development investments that Boeing ever made[4, 5]". He further stated "It was a tremendous improvement. We could have delivered that airplane[5]". He somberly & succinctly summed up the prevailing sentiment within Boeing's 7J7 development team, "We went through the various stages of grief in about five minutes[4]." referring to program's eventual scrapping.

However, the development effort done by Boeing on the 7J7 program paid off in a different way altogether. A number of key technologies pioneered & developed by Boeing for the 7J7 program, which included, integrated avionics, flat-panel display-units, fly-by-wire technology, new materials and crucial operating experience with Computer Aided Design (CAD), all of which found their way into the 777 program ultimately[4] , which began right from where the 7J7 had ended. The 7J7, thus, in a way enabled Boeing to do its pre-work, with reference to development of cutting edge technologies of the time, which effectively shortened the development horizon of the 777 substantially to less than half a decade, something really stupendous for a scratch up wide body aircraft program of that scale & scope. Mulally further added, "We could not have done the 777 without the 7J7;[5]" indicating towards the significance of the 7J7 project for Boeing. "The cool story about the 7J7[5]," Mulally said in a 2005 interview, "is that it's exactly the same [idea] as the 7E7, but at a smaller size[5]" with the 7E7 referring to Boeing's 787 program, which as per Mulally, too had been anchored in some ways on the 7J7 work undertaken by Boeing.

"Judge a man by his questions rather than by his answers"– Voltaire

CHAPTER 4

"THE PROP FAN THEORY" AND THE 'CRITICAL CALL'

"Simplicity is the Ultimate Sophistication" – Leonardo da Vinci

At this moment, it would be interesting to understand the background context of the Propfan technology a little in order to make deeper sense of the prevailing, almost gospel-like fervor prevailing in the commercial aviation industry towards the same while also decoding the reasons for disparity & completely contrasting approaches taken by Airbus & the U.S. based OEMs, Boeing & MD respectively towards Propfans.

Propfans or Open rotor technology was being touted as the next big thing in aviation in the 1980s and a radical, generational leap in the commercial aviation industry's eternal quest for evolution. With its genesis in 1975, right in the middle of the 1970s oil crisis, under a NASA Aircraft Energy Efficiency aeronautical research program aimed at potentially developing highly fuel efficient engine technology for aviation applications by blending the speed & performance of traditional turbofan engines and the fuel efficiency of turboprops to take on the fluctuating vicissitudes of the global crude oil market.

The oil prices in 1986, following the skyrocketing levels of the 1970s, stood at the rock bottom levels of 50-60 cents, which still were almost five times the levels they had been at prior to the 1973's crisis[1]. However, given their unpredictable trajectory, the aviation industry was avidly looking forward to the prospects of almost 30% improvements in fuel efficiency being proposed by the prop fan technology over traditional turbofans.

Propfan technology essentially involved a large propeller fan rotating without being encased by the large, heavy duct with blades rotating at virtually supersonic tip speed to generate very high-bypass ratios of almost 60:1 which were at the core of the huge fuel efficiency gains promised by it. The leading aircraft engine manufacturers in the U.S. had already been pursuing their own R&D programs on propfans at the onset of the 1980s decade, led by GE's Unducted Fan (UDF) & the Allison-Pratt & Whitney JV with its 578-DX, with NASA seeding them with the outcomes of its own preliminary research efforts on the technology.

However, managing in-cabin noise was going to be the biggest hurdle to be overcome before the propfans could make it to the commercialization stage which was expected to be achieved by the industry by 1992. Boeing had gone a step further and was targeting ambitious fuel efficiency gains of around 60% on the propfans powered 7J7, as compared to traditional turbofans, with additional fuel efficiency gains expected to be realized from usage of lightweight materials in the airframe, design & development of high lift/low drag wing and incorporation of an advanced flight control system.

Further, a 1987 NASA document prophetically concluded that "barring other, unforeseen problems, the prop fan airliner could be carrying passengers within five years[1]". This deification of prop fans as the next savior of commercial aviation in a post oil crisis world in the 1980s was not just a regional phenomenon with a multitude of aviation players across continents, including, European players like Fokker, ATR, Aerospatiale & Messerschmitt to the Soviets, led by Tupolev and Ilyushin; having studied or proposed propfans based commercial aircrafts through the 1980s decade.

The Airbus take on Prop fans, however, contrasted completely from its North American counterparts despite the fact that SNECMA, the key French aviation engine manufacturer had already been invested in the development of the UDF technology at the time; under CFM International, its engine manufacturing JV with GE Aviation established in early 1970s; having made a 35% investment in GE's UDF propfan engine program, which later became the GE36. In fact SNECMA's reigning chairman in late 1986 had commented, pitching the tent for the UDF, "The in-development CFM56-5S2 would be the last turbofan created for the CFM56 family, and that there is no point in spending more money on turbofans. UDF is the future[2]". Rolls Royce, the leading British engine manufacturer, too was in the race and had placed its bet on the wing mounted propfan concept, having designated it as Contrafan.

Airbus, which still was a relatively new & relatively lesser known European commercial aircraft OEM in the mid-1980s; when the propfan technology & the theories about its potential to revolutionize air travel were at their peak; had while reviewing the likely evolution of the industry's technological landscape considered prop fan technology while conceiving its latest narrow body aircraft program, the A320.

The A320 was one of the most conventional designs of the times and was the result of ultimate decision by Airbus to use regular, conventional turbofan engines instead of following the herd and developing an exotic propfans powered aircraft of its own. Devoid of the archangelic guiding light & deep R&D foundations of NASA and facing constraints pertaining to resources as well as the development horizon available to get the A320 airborne; Airbus, thus, had to take the most prudent call pertaining to selection of the propulsion system.

In this context, an opinion shared by the then Vice President of for development of new products and advanced technologies at Airbus, D. Little, made at an industry conference in the mid-1980s sheds more light on the Airbus take on the subject. He said: "The use of modern techniques and innovations can be very costly and the question 'whether' the resulting product justifies the effort in terms of profits is equally essential to the aircraft manufacturer and to the operator[3]" referring to the application of advanced technology on future commercial aircraft.

Airbus played it really smart on the A320 program by going for and incorporating proven & mature technologies, like advanced structural materials, fly by wire & advanced avionics which could enable the European air-framer to effectively achieve a generational leap while ensuring profitable utilization of these advanced technologies without endangering the development schedule and entry into service timeline of the program and staying clear of the investment intensive, unproven & risky ones which could have played spoil sport. Key factors based on which the decision was taken by Airbus included:-

1. The development curve of the prop fan technology was going to make it available for commercial usage only by 1992 at the earliest, which was way beyond the entry into service schedule of the A320 slated for 1988.

2. Potential integration issues of the prop fan with the airframe which would have taken additional development time and could have impacted & risked the originally targeted EIS window.

3. Airbus while conceiving the A320 had already packed it with enough advanced & cutting edge technologies of the time, like, fly by wire systems, advanced avionics, enhanced wing aerodynamics, usage of composites, FADEC[3] enabled CFM56-5 & V2500 engines and an automated maintenance system. The incorporation of these technologies was going to make sure that the A320 was indeed a generation ahead of its contemporary, in-production aircrafts of the time, either from Boeing's or McDonnell Douglas' stable. The overall package of the A320 proposition was geared towards ensuring the airlines operator's profitability through the aircraft's in-service operational life span; projected initially in early 1980s to be at least through the year 2010; which was taken as the program's projected service horizon based on initial assessments & review of the market's technological landscape evolution (and Airbus was bang on target given that the A320neo was launched with next-gen engines precisely in December 2010!).

Thus, rather than developing the most technologically advanced aircraft of its time, Airbus focused on operators & its own profitability and left no stone unturned in ensuring this. This was a radical departure from Boeing's approach on the 7J7 program (which consequently did not find any takers ultimately). Airbus' then Chief Planner, Adam Brown, believed that the 7J7 was an attempt by Boeing to disrupt the A320 program with Boeing creating hype around it[4] trying to deter Airbus from pursuing the A320.

Adam Brown refers to the 1985 Paris Air Show where Boeing showcased the 7J7 and posed the existential crisis question to

---

[3] Full Authority Digital Engine Control

Airbus questioning the very foundation of the A320 program & its future compared against an aircraft (7J7) which was claimed to be over 50% more fuel efficient than the A320 would be. "We can go up against the 'magic aeroplane,' and we can beat it[4]." said a confident Adam Brown recalling the prevailing scenario & sentiment in an interview years later.

However, within a year and a half, by early October 1986, Airbus had a firm order backlog of 119 aircrafts along with options for another 144 aircrafts on the A320 program; with the aircraft (a first narrow body from a not so well established European OEM) not even having undertaken its maiden flight and still years away from entering commercial service. This was a big boost for Airbus and must have given Airbus a lot of confidence over its decision to go with conventional turbofans. This order backlog for the A320 had rapidly grown within a span of 20 months to 500 aircrafts on order by June 1988; post the A320's entry into service (EIS) in April 1988.

4. Airbus' revised strategy of penetrating the North American commercial aviation market with the A320 narrow body, post the limited success of the A300/310 wide body aircrafts beyond continental Europe, was premised on providing the airline operators with a low risk capital investment option to try out a relatively less proven but technologically superior European product. Therefore, choosing the unproven propfan technology along with its unsorted, huge noise problems in a relatively low fuel price environment of the 1980s would have meant injecting a significant amount of risks & costs in the program unnecessarily

which would, in turn, have been counterproductive to the Airbus'
original low risk strategy.

5. The A320 program, from the outset & design stage itself was being
    developed as a family of aircrafts with plans for maintaining
    significant commonality across aircraft programs with reference to
    key systems, avionics & engines etc. This strategy was also referred
    to, internally within Airbus, as 'Cross-pollination'[3] with the same
    focused again on ensuring profitability of airline operators & self.

    This approach of maintaining commonality of systems across
    aircraft programs kicked in tremendous scale advantages &
    economics in procurement trickling down to the Airbus bottom
    line. For the airline customers, it meant ease of moving pilots &
    crews from one aircraft to another without incurring any major,
    additional training costs translating into lower operating costs and a
    boost to profitability which also turned into a minor, indirect
    incentive for sticking to the same aircraft maker & type and for
    maintaining fleet commonality which has led to the evolution of a
    multitude of airlines, especially in the Low-Cost Carrier (LCC)
    space, maintaining a single aircraft type across their fleets.

    For instance, Irish LCC Carrier Ryanair, founded in 1984, with
    an all 737 fleet or the Austrian LCC carrier, Laudamotion's all-
    A320 aircraft family fleet. This cross pollination strategy of Airbus
    would have preferred the relatively much more stable and mature
    turbofans (in terms of technological evolution) as the core pivot
    from a long term perspective rather than the emerging prop fans.

6. There were a lot of technical issues pertaining to prop fan
    technology which were yet to be addressed & sorted out before the
    technology could be deemed as mature and ready for

commercialization. The most pertinent issue with the technology pertained to managing cabin noise levels as the ductless fan with its carbon fiber blades rotating at supersonic tip speeds would have created huge noise problems besides creating issues of structural fatigue to the airframe due to vibration. The prop fan technology, in order to be successful, had to effectively bring down the noise levels at least at par with conventional turbofans or to the humanly acceptable levels.

The mounting position of the prop fans was another key point to be pondered upon as mounting them under the wings would have led to in-cabin noise & structural fatigue problems while mounting them on the rear fuselage would have created interference problems with the wing's air flow, fuselage & the exhaust effects of engine nacelles[3].

Further, internal studies by Airbus indicated that aircrafts with aft-mounted engines were heavy with higher maintenance costs. Airbus' chief planner, Adam Brown said: "The answer depended very much on the price of fuel[4]." Additionally, the structural strength & the ability of the propfan's ductless rotating blades to withstand potential bird strikes was questionable, in addition to their vulnerability to Foreign Object Damage (FOD) and water suction problems.

7. There were no existing certification standards in place with the FAA or EASA for propfans which in turn had to be developed from scratch and would have further added to the certification process and entry into service horizon of the propfans powered airplanes.

Propfan development efforts; being spearheaded at the time by the GE's gearless UDF technology (which created the GE36 prop fan engine) and the Allison's more conventional geared 578-DX engine; continued & ultimately reached the test flights stage. The Allison's propfan concept was mounted on a modified Gulfstream II aircraft & was tested through 1987-1988. The GE36, developed with 35% participation by SNECMA (now Safran), was first flight tested on a 727-100 test bed in 1986. McDonnell Douglas, which had become the chief patron of the propfan technology development effort, mounted the GE36 engine on a modified MD-80 demonstrator aircraft, which was also showcased at the 1988 Farnborough Airshow.

GE claimed that its UDF technology could cover a broad power output range and was capable of meeting or surpassing the power output of its CF6 turbofans which were the prevailing gold standard amongst wide body aircraft engines at the time. The GE36 was planned to be developed in four variants with thrust output ranging from 14,000 lbf for the MD-91X to 25,000 lbf for the MD-92 & 7J7 programs while a 22,000 lbf variant was to power the MD-91. McDonnell Douglas also tested the Allison's 501-M78 engine on its MD-80 test bed in 1989.

Image 1: Model of the Unducted Fan (UDF) Engine on Display at the Safran Museum

Image Credits: Duch, Wikimedia Commons, CC-A–SA 4.0 International

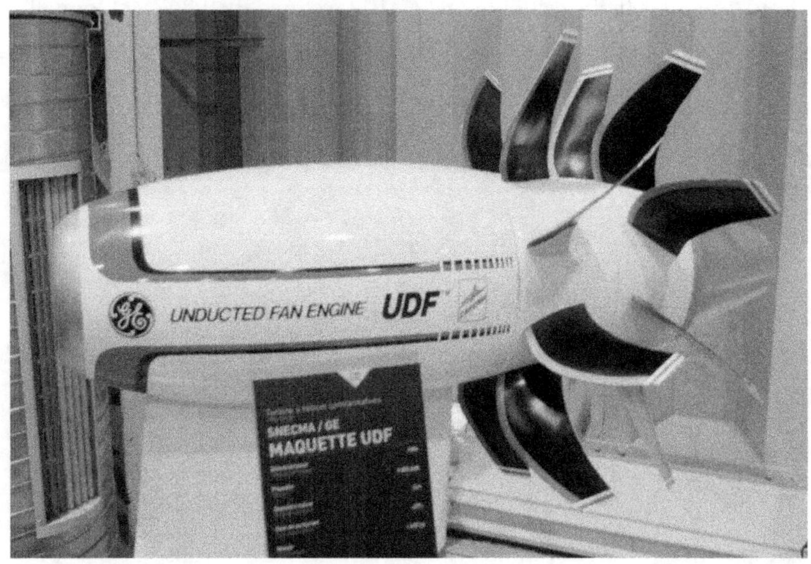

<u>Image 2</u>: The GE36 on a McDonnell Douglas MD-81 (UHB) demonstrator at the 1988 Farnborough Air Show.

Image Credits: Andrew Thomas, Wikimedia Commons, <u>CC BY-SA 2.0</u>

However, the propfan technology ultimately never saw the light of commercial service day with the tracks of the path leading to the future usually determined & laid by indifferent & impersonal market forces driven by the underlying economics and their verdict at the time was against it.

The cabin noise problem & prevailing low oil price environment questioned their very viability. Further, all market players involved in the development efforts faced different set of issues which also impeded any further progress on the commercialization of propfans. Boeing wound up the 7J7 program without any traces of it resurfacing ever.

McDonnell Douglas (MD); having launched the MD-90(derived from

MD-80) single aisle airliner powered by the V2500 turbofan in 1989 to take on the A320; too, abandoned the MD-91X, MD-92X & MD-94X programs altogether amid shifting priorities. Al Novick, who was part of the Allison-P&W team that had developed the 501-M78 engine, read out the program's cryptic obituary: "The operation was successful, but the patient died[4]" referring to MD's abandoning of the development of propfan powered aircraft programs post the successful flight of the MD-80 test bed powered by the 501-M78 engine in 1989.

GE, which had developed the GE36, was initially touting the GE36 as a replacement for the now iconic CFM56 turbofan which had been faring poorly in terms of sales uptake from airlines during its initial half a decade of introduction in the mid-1970s. However, towards the late 1980s the CFM56 program had gained a lot of sales momentum helped partially by the technical issues faced by the V2500 and GE, thus, did not want the program to be cannibalized by the new kid on the block, the GE36. GE, much akin to Boeing, leveraged the UDF development effort & ultimately injected the UDF's fan blade technology into the GE90 engine program, which eventually found its way into powering the Boeing's 777.

For Allison, the benefits were not direct, recalls Al Novick, saying that "We got involved publicly with Boeing and Douglas and it helped us tremendously in getting back into the commercial business[4]" referring to the fact that the Allison turbofan AE3007 was later selected by Embraer for its EMB-145 regional jet program & Allison ultimately was acquired by Rolls Royce. Alan Mulally, Boeing's Director of Engineering for the 7J7 program summarized the UDF story in a nutshell in an interview much later as: "The biggest issue with the UDF was to make it a simple engine and get the reliability up and the maintenance down[4]".

Further, the onset of a radical shift in traditional world order, with the official ending of the Cold War led to the onset of a tumultuous decade for the Aerospace & Defense industry in the United States and most parts of the world marked by virtual evaporation of defense spending & budgetary allocations towards R&D on next generation technologies besides triggering tremendous & unprecedented industry consolidation that witnessed virtual disappearance of a number of top of the line defense primes, which were virtually household names, from the industry radar. McDonnell Douglas, the third key force in commercial aviation at the time, besides Boeing & Airbus, was ultimately acquired by Boeing in 1997.

This defense spending meltdown closed all possibilities of any potential military application of the propfan technology as well. McDonnell Douglas, in fact had proposed a DC-9 based prop fan powered maritime, anti-submarine patrol aircraft to the U.S. Navy, for its LRAACA program, as a potential replacement for the Navy's Lockheed built P-3 Orion fleet. But the Navy chose the competing Lockheed P-7A offering just prior to the LRAACA program's deferment as it got itself stuck into the tightening firewall surrounding defense spending at the time only to resurface much later as today's Boeing built & 737 based P-8 Poseidon.

Propfan saga is more or less over, despite the techies love for it or CFM International claiming the Open Rotor technology as the future of engines during the 2000s decade given the much stricter noise regulations in place now and the onset of the industry's shift towards hybrid & electric propulsion geared towards future driven by significant challenges & pressures emanating from the sustainability front on aviation. The Prop fan hype, somehow, was completely in line with the Gartner's Hype Cycle with the 'Peak of Inflated Expectations' followed by the 'Trough of Disillusionment' perfectly!

Airbus, thus by choosing to stick to turbofans, got a difficult, important & critical call right for itself basing it on a mix of assessments grounded deeply in reality & common sense despite Boeing's marketing gimmicks & might; which along with the permanent grounding of the Boeing's 7J7 program; provided the perfect launch conditions & runway for the A320 to become airborne and this was precisely the moment in time & history which marked the onset of a shift in the gravity & strategic calculus of the global narrow body aircraft market from North America towards Europe from the supply side perspective.

The trend was further accentuated by the radical & sudden shift in Boeing's strategic focus towards the wide body aircraft market from 1990s decade onwards, riding on the globalization wave, which positioned Airbus well & stably on the trajectory to ultimately be able to be in a position to almost wrest the crown of the narrow body aircraft market from Boeing almost two decades later. Boeing's strategy for the narrow body aircraft market from hereon could be described as reactive at best and complacent at worst.

Squandering the Golden Window of Opportunity

The key window available to Boeing to create an effective replacement for the 737 or to get a slightly larger, clean sheet, next generation narrow body flying workhorse into its stable and the market; which could have doubled up as & could have led to the creation of a further stretched, derivative single aisle aircraft targeted at the Middle of the Market; and which could have effectively taken on the might of Airbus A321 and its longer variants, LR & XLR, in the 21st Century.

For Boeing, the scrapping of the 7J7 meant losing a great window of opportunity for creating a new, clean sheet design based scratch up, narrow body aircraft which could have effectively taken on the technological edge of the A320, the comparative advantages of which (against the 737) were soon experienced as well as acknowledged by the industry, especially with the fly by wire controls & flight envelope protection.

"Boldness governed by Superior Intellect is the mark of a Hero". – Carl Von Clausewitz

CHAPTER 5

THE 'STRATEGY SHIFT' THAT TRIGGERED A 'TECTONIC SHIFT' AND BROUGHT 'TWO JUMBOS' DOWN

"Everything in War is Simple. But the Simplest Thing is Difficult". – Carl Von Clausewitz

Boeing's product strategy in the 1990s had radically shifted its focus on to the wide body aircraft segment with Boeing's internal market forecasts indicating towards a virtual windfall for commercial aircraft manufacturers over medium term with significant demand projected for wide bodies driven by the onset of forces of globalization from the 1990s decade onwards which were projected to drive a significant surge in demand for intercontinental air travel. Boeing's market forecast for commercial aircrafts released in 1989 projected doubling of aircraft traffic over a 16 year period spanning 1989-2005, with air traffic projected to clock a growth rate of 5.9% annually over the 1989-2000 period & 5.4% subsequently through 2005[1].

This was likely to provide significant tailwinds with projected demand for over 8,500 new aircrafts through 2005 with almost 70% of them likely to be ordered by airlines for fleet expansion & growth with the remaining 30% for replacement[1]. However, analysts differed & termed Boeing's projections towards the lower side and pegged the replacement demand to be higher than that. George Shapiro, a stock analyst at the time with Salomon Brothers based in New York commented "It seems like more than 30 percent of the upcoming new orders would be for replacement, considering that in another three years about 3,000 jets will be 20 years old,"[1] indicating towards almost an upcoming gold rush in the wild west setting with a potential market opportunity worth almost $420 billion.

Boeing was positioned extremely well to capitalize on this upcoming tailwind as the incumbent with an almost 60% market share at the time. Boeing had an order backlog of around 1,000 aircrafts at the time and had delivered 269 aircrafts worth almost $10 billion in 1988, as per industry estimates[1].

Boeing's forecast clearly indicated towards an emerging shift in demand towards wide bodies after 1995. The report said: "larger jets, such as Boeing's 747 and 767, will be in higher demand after 1995. There will be a shift to 57 percent of the seats being in jets with more than 350 seats[1]". The forecast projected 50% of the income to come from the larger, wide body planes owing to lower seat costs to airlines & added benefit of airport decongestion.

Boeing's report concluded by forecasting demand for over 2,356 large wide body airplanes worth $251.3 billion[1] likely to account for almost 60% of the total airplane market in value terms through 2005 with the 120-170 seat single aisles projected to lead in volume terms, a trend which stays on track to this day with single aisles consistently leading volumes while the wide bodies dominate in value terms.

Additionally, as per Boeing, the Intra-Europe air travel market was projected to grow at the fastest pace of 5.8% followed by 4.9% for the U.S. market, which would surely have pleased & thrilled Airbus given the A320 had just entered service in 1988!

Armed with the knowledge of this emerging growth trend and a readjusted compass; Boeing recalibrated its overall product strategy, going for the hunt back to its favorite hunting ground, the wide body landscape from thereon for the next two decades starting with the creation of the extremely successful clean sheet 777 in a record time during the first half of the 1990s followed by the development of the 747-8 in mid-2000s while dabbling with the point to point transportation model in the late 1990s with the proposed Sonic Cruiser concept, unveiled in 2001 and sinking without a trace and later re-emerging on to the surface, reincarnated as the 7E7 concept, becoming the 787 ultimately.

Airbus had, in fact, already set course for that journey having already launched the A330 & the A340 as the evolutionary pathway for the A300/310 in 1987 and the legendary battles between the A330/340 and the 777 were thus scheduled for the 1990s decade, to be shaped & adjudicated eventually by the ETOPS regulations, which eventually ruled in favor of the 777 and drove its tremendous commercial success.

Boeing's crusade in the narrow body aircraft market, in the meanwhile, was being spearheaded by the same old legendary warrior, the 737 at the core (with Boeing having missed the bus on the 7J7 effectively) which Boeing was soon going to be cranking out in huge numbers with the program's third avatar, the 737NG, already on its way. The 737, however, was clearly ageing having already seen two generations of aerial warfare with Boeing having no plans in sight or on the drawing board for a potential replacement post the 7J7 debacle.

Boeing's strategy for the narrow body aircraft segment, thus, could be described as reactive at best highlighted & underscored by multiple product line extensions employed at the eleventh hour as a knee jerk reaction to competition's tactical moves till the old workhorse finally broke down much later while exposing the skeletons hidden in Boeing's closet. The question to be asked here is did Boeing push the 737 airframe beyond its limitations despite having generated huge profits for decades based on the pursuit of its low cost derivative strategy on the 737 program?

Boeing's ultimate response to the A320 in the early 1990s, in form of the launch of the third generation of the 737, the 737NG in 1993, is somehow up for debate from a product strategy perspective. Though, the same has been a huge success from the perspective of bean counters as well as in the economic sense overall, with over 7,000+[2] 737NG aircrafts ultimately rolled out by Boeing over the program's 22 year long production run through late 2019.

The 737NG, in fact, has been the most successful generation of the 737 program in the almost half a century long saga. This is without including the units produced for the 737 airframe based military variants, including, P-8 Poseidon, the maritime patrol & anti-submarine aircraft developed for the U.S. Navy to replace the P-3 Orion fleet, the 737 based AEW&C variant built for some international customers and as the C-40 Clipper created for the U.S. Air Force for personnel transportation. However, this was the over two and a half decades old, 1960s era 737 airframe on steroids (737NG), while also incorporating some of the technologies developed for the 7J7 and not a clean sheet aircraft built from scratch with structural airframe limitations very much intact.

A Foible which ultimately would become the Achilles' Heel

One of the old bottlenecks was again confronted by Boeing while mounting the new CFM56-7B engines on the 737NG's low-mounted wings in the early to mid 1990s which was tricky and required a tedious engineering effort on the part of CFM International this time as well, including, changing the engine nacelle shape from round to oval once again and adjusting the power output rating accordingly. This was a déjà vu moment for the 737 program with the 737 Classic having faced the same challenge in early 1980s with the mounting of the CFM56-3 engines under the wings.

The 737's legacy of low mounted wing configuration dates back to the commercial aviation world of 1960s and was geared towards efficiency in an era of manual ground operations. The 737 was designed primarily for short, regional operations wherein flight frequencies and daily cycles have a relatively much more direct bearing on the airlines bottom line. The low mounting of the wings facilitated typical ground crew operations ergonomically in the 1960s era, including, manual baggage handling & catering services, faster enplaning & deplaning for passengers using folding stairs and easier access for maintenance personnel for carrying out on-wing engine maintenance activities, all of which in turn translated into quicker turnaround times for the airlines.

Image1: A 737-200 belonging to the Aerolíneas Argentinas (below)

Markus Hening, Wikimedia Commons, Public Domain (PD)

Image 2: An Air North 737-201 (below) Makaristos, Wikimedia Commons, Public Domain

Further, the 737 was originally designed by Jack Steiner & Joe Sutter with engines initially positioned on the aft rear fuselage, much like the 727, rather than under the wings. The first pair of Pratt & Whitney low bypass JT8D turbofan engines, which the 737 Original sported, were essentially meant for that with their cylindrical shape & cigar like nacelle design (see image 2 above) intended for fuselage mounting.

The JT8D engines were developed by Pratt & Whitney for the 727 in 1963 and were ultimately used mostly on narrow body aircraft programs with fuselage mounted engine designs, led by the 727, DC-9 and the MD-80. The 737 was the only aircraft program to have mounted the JT8Ds under the wings and it was Joe Sutter; prior to his shot to fame with the 747; who re-jigged the 737 design at a later stage and instead repositioned the engines by mounting them under the wings[3] for weight and structural stability reasons which also enabled him to widen the fuselage a little more for six abreast seating.

The ground clearance was never an issue for the original pair of 737-

100s & -200s with the cylindrical JT8Ds but later when it came to re-engining with larger & conventional engine nacelle shapes, it caused problems and in fact became a structural limitation but still was never addressed by Boeing to keep the same type certificate across 737 generations. It was somewhat of a foible for the 737 which when pushed to the limits by Boeing ultimately would become its Achilles' heel as the program as was never conceived & designed in the first place for an operational service horizon spanning over half a century!

However, thanks to 'Visionary Joe' for his late cognitive spark in mounting the engines under the wings which enabled Boeing to be able to extend the shelf life of the 737 by being in a position to re-engine the 737 multiple times over the subsequent decades, unlike fuselage mounting, which would have surely impacted the longevity of the 737 by having compounded the underlying range & complexity of problems, issues & limitations for Boeing many times over while considering & attempting re-engining later!

This was a structural limitation and Boeing had to be aware of the fact that it was going to face the same issue again in future if the 737 needed to be re-engined rather than outright replaced, given the engine technology evolves every 15 years on average, which is what happened. Thus, in the economic sense, the decision turned out to be a clear success but while also raising a question on the 737's remaining life span & succession, given re-engining constraints, as Boeing couldn't have used the 737 airframe forever.

Further, the engine technology has been moving upwards towards achieving further higher bypass ratios for achieving an incremental increase in fuel efficiency and was likely to produce larger and heavier engines in future as well which would have further compounded the re-engining problems for 737 going forward, if the need arose.

Image 3 (Below): A CFM56-3 Engine Mounted on a Japan TransOcean Airlines 737-400; Wikimedia Commons, CC BY-SA 3.0

Image 4 (below): A CFM56-5 Engine on an A320: (Compare the difference in Engine shapes): Curimedia, Wikimedia Commons, CC BY 2.0

With the scrapping of the 7J7, Boeing effectively lost the golden window of opportunity it had to replace the 737 or to add a brand new narrow body aircraft, custom-built for future to its portfolio which could most importantly have been leveraged by Boeing to solve two of the very pertinent product strategy issues it was going to face two decades later. First, Boeing could instead have used this new 7J7 based airframe for re-engining in the early 2010s rather than creating the MAX by re-engining the old 737 and pushing it to the hilt. Secondly, larger stretches of this new aircraft could potentially have been positioned & used very effectively by Boeing to tackle Airbus' current stranglehold on the middle of the market led by its further stretches of the A321 to produce the A321LR & the A321XLR. However, that's hindsight and it's always 20/20!

Boeing revisited the drawing board once again, considering a 737 replacement in the mid-2000s, while it was involved deeply in the production of the first clean sheet aircraft program of the 21st century, the 787. This time the aircraft project was code named Y1, which was part of a comprehensive, sweeping modernization program aiming at replacement of Boeing's entire commercial aircraft portfolio with newer, clean sheet aircrafts incorporating cutting edge, next generation technologies. The program was codenamed 'Project Yellowstone' or 20XX by Boeing.

Project Yellowstone/Project 20XX

Project Yellowstone, also referred to as Project 20XX, was an ambitious program thought out by Boeing's leadership at the turn of the century aimed at renewing Boeing's entire existing civil aircraft product portfolio by replacing it with just three next generation aircraft platforms incorporating cutting edge technologies. The game plan proposed creation of three distinct, clean sheet aircraft programs, code named, Y1, Y2 and Y3, which were to be custom built & positioned to cover the entire gamut of commercial aviation market segments while bringing in tremendous advantages of scale, scope & commonality to Boeing.

The plan was geared towards making the company able to compete much more effectively against Airbus on total cost of ownership by offering attractive pricing deals to airline customers on Boeing airplanes along with offering much lower operating & maintenance costs emanating from utilization of common systems & aircraft parts through the entire life span of aircrafts. Boeing had planned to create & leverage commonality in design, systems, operations as well as manufacturing of these aircrafts aiming for a simpler, common aircraft fleet, much akin to the approach taken by Airbus while conceiving its A320 aircraft family in the early 1980s.

The new, proposed aircraft line-up to be created would have comprised 3 aircraft platforms as against 6 distinct aircraft models which formed Boeing's commercial airplane portfolio at the time.
Yellowstone, in turn, referred to the technologies which were being focused upon by Boeing for the program which ranged from composites based aero-structures, incorporation of electrical systems instead of the traditional hydraulic ones and next generation turbofan engines (which had already made their appearance felt by knocking on the industry door during the early 2000s led most prominently by the Pratt & Whitney's geared turbofan engine technology based PW1000G engine family).

Boeing, thus, was in essence trying to replicate the success of the Airbus' A320 aircraft family strategy of the 1980s in the 21st century; looking to undermine this significant Airbus advantage, while also extending it across the board aiming to cover its entire aircraft portfolio. The strategy also involved dovetailing the plan with the ongoing waves of globalization through the pursuit of an outsourcing heavy aircraft manufacturing model, pivoted on shared development costs, with a key role reserved for the Japanese aviation industry heavyweight samurais in the script, as partners as well as customers. "If we pull this off, we'll ruin Airbus. We'll reinvent the whole business,"[4] said one senior technical employee on the plan. He further said "But if Boeing doesn't build the 7E7 in Washington, they'll never build another plane here[4]" referring to the fact that the location chosen by Boeing for the first aircraft program would in turn was going to be critical and would be leveraged to produce other members of the proposed aircraft family going forward. 7E7 was the Y2 leg of the Yellowstone which was announced in January 2003 and launched ultimately as the 787 in April 2004.

The Project 20XX was a derivative of Boeing's another plan which was first outlined in the late 1990s after having significantly exceeded the original development budget on the 777 program. The plan was termed as 'Aircraft Creation Process Strategy' leaning heavily towards aircraft family, derivatives, commonality and technology harvesting concepts.

The plan was geared towards drastically cutting down on the huge upfront investments involved in new, clean sheet aircraft programs along with cutting back on the traditional development horizon of almost a decade. The ambitious goal was referred to by Boeing as 'One in Ten' translating into keeping the 'upfront investments to around $1 billion and a development horizon of 10 months from design to first flight'[4]. This Boeing master plan of taking a radical approach on its next clean sheet aircraft development program ultimately precipitated into the $32 billion financial disaster, designated ultimately as the 787 Dreamliner!

The 3 aircraft platforms being conceptualized by Boeing under the program included:-

Boeing Y1: The Y1 was targeted at the lowest end of the Boeing's aircraft portfolio covering the single aisle and middle of the market segments with a passenger capacity of 100-250 seats and was being conceptualized as a potential replacement for the existing 737, 757 & 767-200.

Boeing Y2: The Y2 was geared towards replacing the 767-300 & -400 aircraft models aiming for the lower end of the wide body aircraft market with a planned capacity of 250-350 passengers. The Y2 was to take on the Airbus wide bodies, namely, the A330 & A340 and the yet to emerge in response, the A350.

Boeing Y3: The Y3 was Boeing's dart for the wide body aircraft market, covering the 350-600+ passenger seat aircraft segment, as an ultimate replacement for the 777-300 & the iconic 747. The Y3 was to take on Airbus' wide body escapade, the A380 and the A350XWB later.

Arch-rival Airbus, however, was looking to execute & materialize its already brewing up plans at the turn of the century to counter Boeing's traditional domination of the long haul market with the 747. Airbus played it much like a pride game, amid the backdrop of the prevailing intra-European political fissures & factions, which had a strong clout & influence within Airbus.

The Europeans, looking for their version of the Joe Sutter designer masterpiece to showcase and as their contender in the battle of the aerial heavyweights, launched the A380 program to counter 747 in December 2000, duly supported in the process by Emirates, with an initial development budget of €9.5 billion marking one of the biggest strategic errors from the European plane maker in its history, in the process, with the A380 program never really taking-off in the economic sense with only 242 A380s built & delivered by Airbus by 2019, over almost 2 decades of program's operational life span. Compare that to the 1500+ 747s produced & delivered by Boeing in almost over 5 decades since EIS in 1970.

These (the onset of the 21st century) were difficult times for commercial aviation, especially in the U.S. with the unprecedented 9/11 terrorist attacks having wreaked havoc on it sending the market into a literal tailspin followed by a period of marked downswing underscored by significant industry consolidation.

Boeing's internal market research at this time was indicating towards the market shifting from the traditional hub and spoke air transportation model which had prevailed for decades, sustained by the larger wide body aircrafts like the 747, towards more point to point, direct travel between city pairs amid the palmy decades of globalization.

Boeing; guided by the insight, determined to crush the A380 and taking full advantage of being in the reactive strategic space by playing its cards late as Airbus had already committed itself to the A380; having cumulatively spent a sum of EUR 4.6 billion in R&D as well as CAPEX on the A380 program over FY2001-FY2003; accordingly took the 'right' turn at the emerging fork in the road by launching the 787 program, codenamed Y2 or 7E7, in April 2004 after having nipped the outlandish sonic cruiser project in the bud a little earlier.

The 787 went on to ultimately & virtually make the hub and spoke model redundant by enabling airlines to open multiple, new city pairs through direct flights & thus reshaping long haul aviation fundamentally. This success, however, turned out to be a real expensive one for Boeing, pegged at almost $32 billion!

The A380, on the contrary, struggled throughout its innings in terms of sales tally with Airbus announcing in February 2019 that it will pull the plug on the A380 program with production of the A380 scheduled to be shut down in 2021 given softening of market demand & Emirates backing out of it, after a short & limited production run without the program having reached the break-even point, with Airbus having produced a mere total of 242 A380s as of December 2019.

The A380 clearly was an aircraft behind its time (at least by a decade) and was a program that had run its course with Airbus unable to recoup its huge investments in the program even from secondary applications, like freighters, given the 747's & 777's dominance of the freighter market effectively denying the A380 any potential easy foray. The A380 program never really took off in the freighter role despite its huge cargo carrying capacity, the largest for any commercial freighter aircraft, despite Airbus having considered & made efforts for creating an A380F.

The era of the jumbos, the extremely large passenger aircrafts, the sort of 'airborne cruise liners', seems to be finally coming to an end in the post COVID-19 altered skyline with Boeing also announcing the ending of the production run of its iconic 747 program from 2022, after a strong, half a century long successful run, with the announcement towards the eventual abdication of the throne by the 'queen of the skies' made by Boeing in late July 2020, as part of the company's H1 2020 financial results. The COVID-19, in fact, has broadly & significantly accelerated the process of retirement of quad jets from airlines fleets across most markets & regions, thus, effectively setting the sun on these mighty, fuel thirsty four engine flying beasts belonging to a bygone era!

With the pursuit of the 787, Boeing tried to create a double body blow for Airbus. First, the 787 effectively decimated the market prospects for the A380 by unlocking the potential of the point to point model. Second, the creation of 787 unleashed & ratcheted up the pressure further on Airbus from its traditional airline customers; especially the Middle East based carriers, which actively sought an Airbus' salvo to counter the formidable, composites based 787 with the A330 being no match for it; which forced

Airbus to ultimately put in another EUR 11 billion reluctantly into the development of the clean sheet A350XWB program in 2006 while the A380 was still taxiing on its way for the take off roll.

This was significant given that Airbus had already spent EUR 10 billion on the A380 program in almost over half a decade spanning FY2001 to FY2006. Additionally, Airbus had an overall negative impact of EUR 3.5 billion on its EBIT for the FY2006 emanating from delays on the A380 program apart from issues on the A400M & A350 programs leading up to the launch of the A350XWB program in December 2006.

Boeing indirectly also tried blockading & cutting off Airbus' financial supply lines by vociferously raising the subsidies issue at the WTO (by the U.S.) launching an inveigh over launch subsidies provided to Airbus by European states in 2004 which led to WTO's initiation of twin probes against both the U.S. & the EU in 2005. With this double whammy; Boeing further upped the ante on Airbus in the wide body market.

However, it wasn't akin to a checkered flag for Boeing by any means either. In terms of numbers, Airbus had invested in the range of EUR 18-25 billion as sunk costs on the A380 without being able to recoup it while Boeing's success on the 787 came at a colossal almost $32 billion in development costs with Boeing still quite a long distance away from the break-even point, facing profitability issues on the program and many of its industrial partners, led by the Japanese heavies, having lost money on the program. Further, by opening the point to point model, the 787 sounded the death knell not just for the A380 but also dug the grave for its 747-8I, the latest passenger variant of the 747.

The stretched 747-8 was the subsequent maneuver of Boeing's overall battle plan, which was preceded by the launch of 787, turning Boeing's campaign effectively into the classic pincer movement aimed at lethal, double envelopment of the Airbus A380 and effectively trapping it. The 747-8I was conceived to leverage the technological advances made on the 787; especially, with the high-bypass ratio GEnx-2B engines, raked wingtips & advanced flight deck; launched in late 2005 and entered service in March 2011 with a redesigned wing & GEnx engines; right after the 787 had more or less annihilated the hub and spoke model; that had originally led to the genesis of and had been the very 'raison d'être' of the jumbos.

The 747-8 program; launched by Boeing with an initial 300 aircraft production run target split across passenger (747-8I) & freighter (747-8F) variants; consequently, has seen the 747-8I becoming the least selling 747 passenger variant ever with only 47 orders received by Boeing since launch to date with the last one coming in 2017.

The damage to the 747-8I's overall market prospects also included cannibalization by the other modern, wide body twin in Boeing's portfolio, most notably, the 777-300ER; which entered service in 2004 and almost replaced the 747-400. The 777-300ER, in fact, had been the Boeing's most successful wide body aircraft variant in terms of sales numbers till it was overtaken by the 787-9. The 787, thus, created significant collateral damage in the process of slaying the A380 and is, in turn & in a way, also a convict charged with Airplane fratricide!

Many industry analysts also have seriously questioned the ability of the 787 program to turn a profit ever even after the order book standing past the 1500+ orders milestone by June 2020. An economic model by International Institute for Strategic Leadership (IISL) projected that even after having delivered 2,000 787 Dreamliners, Boeing would still incur "a total program loss of approximately $5 billion[5]" given the colossal, cumulative deferred costs base. Another industry analyst, Bjorn Fehrm, with Leeham News & Analysis, has developed detailed models on 787 costs and revenue and he said: "No Way. They need to recover far too much money per airplane. They will never do that[5]". Compare the 787 to the A350 program which Airbus was forced to undertake in response to the Boeing's 787 maneuver. As per Airbus, it achieved breakeven on the A350 in 2019.

All in all, the bloody aerial warfare between the two mighty rivals witnessed the launch of a salvo of four wide body aircraft programs by the duopoly at around the turn of the century through the first decade at a collective cost of at least $75-$80 billion. Of those four, two of them (A380 & 747-8) are sure to be ending up in losses with much shorter production runs than expected originally, one (787) is battling the sunk cost fallacy while only one of them (A350XWB) has managed to reach breakeven by the end of 2019.

In a nutshell, first, warfare & especially aerial warfare is highly expensive indeed and is fraught with serious risks of incurring collateral damage and second, complex is the math of commercial aircraft programs with market forces coupled with market timing calling most of the shots & ultimately passing the verdicts with sometimes no clear winners at all!

"The very same desert, a tactician's virtual paradise, is simultaneously also a logistician's worst nightmare" – General von Ravenstein, German Army

Chapter 6

THE 'LEGENDARY WARRIOR' THAT HAD 'NO SUCCESSOR' AND THE CHANGING AVIATION ZEITGEIST

"The Enemy of a Good Plan is the Dream of a Perfect Plan". – Carl Von Clausewitz

Boeing's pursuit of the 787 program & the simultaneous development of the largest 747 variant, the 747-8; effectively meant the 737 replacement was going to go on to the back-burner, at least for the time being. Boeing while making the choice for the Y2 also stated that the Y1 would be picked up after Y2 since the next generation engine technology hadn't yet hit the market and was expected to do so around 2015.

On this count, Boeing was bang on target with Pratt & Whitney's PW1000G PurePower® GTF engine family having ultimately received certification for its first member in early 2013 and entered active commercial service only in 2016. The story has been similar for the CFM International's competing LEAP engine family which has created the next battle of engines in the narrow body aircraft market after the fabled CFM56 vs. V2500 rivalry of the mid-1980s which was well sustained & preserved for over two plus decades till almost the end of the first decade of the current century.

Boeing, however, continued maneuvering to cover additional ground through the 2000s decade with its loyal & fiery steed, the 737, which had been turning almost 40 by now and had come to its rescue as the savior on multiple occasions when the company had desperately needed a warrior, in its perennial warfare for supremacy against the European arch-rival. By mid-2000s the 737 had reached two key program milestones with 5,000 aircraft deliveries and a backlog of over 6,000 aircrafts in the order book with the versatile 737 juggernaut still showing no signs of slowing down.

Key tactical moves made by Boeing with the 737 included, the introduction of additional variants, especially the introduction of extended range variants, the 737-900ER, in July 2005 followed by the 737-700ER in 2006; besides the development of military variants, the P-8 Poseidon, Australian AEW&C Wedgetail program based on 737-700ER (which was later exported to other international customers as well) and the C-40 Clipper, the USAF's personnel transportation variant and business aviation applications as Boeing Business Jet (BBJ) portfolio.

Given such huge strides made by the 737, which made it a virtual 'aviation timeless classic,' it could easily be termed as Boeing's stalwart & a legend in its own right in Boeing's aircraft line-up along with the 777 holding the wide body crown at the top end of the spectrum.

This explains Boeing's somewhat quixotic belief in its immortality as well as invincibility and the subsequent reluctance towards planning for decommissioning & seeking a 737 worthy successor proactively as it would have signaled the beginning of the end for the iconic program to the market, which had been doing fantastically well, especially with the NG generation, even after 4 decades of being in-service and doing so potentially could have jeopardized the sales momentum for the program which had been cruising in top gear at the time.

Boeing, with the launch of the 737-900ER variant in July 2005, extended the 737 line-up further as part of its efforts to take on the A320 aircraft family more effectively. The 737-900ER was a longer range, large capacity variant of the 737-900NG featuring auxiliary fuel tanks for extended range along with strengthened landing gear & other structural strengthening for withstanding additional weight. The -900ER had a seating capacity of 215 passengers in a single class seating layout and an operating range of around 3,200 nmi with 180 passengers[1]. The 737-900ER derivative was conceived & developed by Boeing as the original 737-900 did not get a warm reception from the airlines & was slow to take-off in sales numbers. Additionally, it was developed to take on the Airbus A321 head-on, which was launched in 1994 as the first A320 derivative with a passenger capacity of 185-236 passengers & an operating range of 3,200 nmi with 185 passengers (typical 2-class seating layout) on board[1].

Further, the 737-900ER was positioned by Boeing to cover the middle of the market segment partially, hitherto spearheaded by the ageing 757-200, with the 757 program having entered service in 1983 and removed from the Boeing assembly lines in 2004 with end of production run. Boeing ultimately built & delivered around 500 737-900ERs (as of May 2019) while the A321 program has seen over 1,667 A321-200 aircrafts delivered by February 2020, as per Airbus.

This battle for fielding the best solution for 757-200 replacement amongst Airbus & Boeing was going to be replicated again in the next decade with Boeing falling short on the number of arrows in its quiver in the post 737NG world which was going to make the critical difference.

Boeing, simultaneously, also reluctantly carried on with its succession planning efforts for the 737 program behind the scenes through the 2000s decade with the most serious one glimpsing through in late 2000s in a Boeing patent filed in November 2009[2]. The patent covered an elliptical composite fuselage design in a seven abreast, twin aisle configuration, which later was going to form the base for the NMA program as well, with Boeing looking to leverage its learning in working with composites on the 787.

The twin aisle configuration for 737 sized airplanes was being considered by Boeing as a potential game changer (also an indelible remnant from the 7J7 conception) with the shorter enplaning & deplaning times on a twin aisle configuration likely to enhance overall aircraft utilization rates for airlines and raising the top line in turn. An article appearing in Flight Global in February 2011 mentioned that Boeing CEO Jim McNerney had even given his green flag to an all new, clean sheet narrow body to replace the 737[3]. But the competitive pressure from American Airlines; which had almost allocated its parking slots to the Airbus A320neo; forced Boeing to shorten its market response time by almost half from a decade for an all new aircraft to a re-engined 737 MAX by August 2011 (to be covered in detail later).

In late 2014; after the re-engined 737 MAX had already been under development; McNerney again proclaimed a charge for an all new Boeing composite airplane powered by next generation engines to replace the 737 MAX by 2030 to take on new, clean sheet narrow bodies coming from new international players entering the market, led by China & Russia. He also hinted at going for a bit larger size than the 737 which would have pegged it close to somewhere in the middle of the market terrain. He actually was referring to & laying the foundations for the NMA (New Mid-Market Airplane) instead, towards which Boeing had started leaning towards by the end of 2014 and once again leaving the 737's ultimate replacement puzzle in abeyance (more on it later).

The other key point to be noted here is the radical change in Boeing's overall strategic posture towards new programs post 787, as underscored by McNerney's famous exclamation. He had said: "No more moon shots"[4] referring to clear avoidance of long development horizons on new, clean sheet airplanes spanning decades by harvesting & harnessing developed technologies and their cross utilization across programs rather than starting from scratch, reflecting the painfully long & highly expensive development process of the 787 which had encountered serious budget overruns & extension of timelines.

Boeing's decision to ride the coming growth wave in early 2000s; by choosing to go on the offensive to produce the 787 at the turn of the century (while also looking to neutralize & settle the A380 threat in the process) instead of addressing the strategic imperative of replacing the ageing 737 & the 757 programs in a planned manner proactively; had in turn led to Boeing missing another key window of opportunity on 737 replacement, notwithstanding the phenomenal market success of the 737NG at the time.

Boeing's missing the bus on 737 replacement plan on more than few occasions actually put the company in a reactive strategic space in the narrow body context which has clearly been a leitmotif in the decades long 737 tale and enabled Airbus to seize the strategic initiative & keep it towards its side of the board once the A320 got airborne in the late 1980s.

The reactive strategic space is an alluring zone for many, given that it provides precious, additional time to play it late and after others have committed to their moves. This is what had enabled Boeing to undermine the A380 strategically when it played its 787 card after Airbus had committed itself sufficiently to and beyond the point of no return on the A380 program. However, the problem with the reactive space is that the one operating in it for long becomes predictable and if you keep operating in it for too long sometimes you run out of options without additional time being available to create new ones as one has to catch up with competition in timing as well and when you try to create options by playing it out of turn sometimes you mess it all up terribly and that is what was going to happen with Boeing as well.

Overall, both Airbus & Boeing had their fair share of strategic errors & tactical mistakes while navigating their ways through the ever evolving commercial aviation market landscape. However, broadly & crudely, Airbus errors mostly were concentrated in the larger, wide body aircraft segment, most noticeably the creation of the A340 as a quadjet rather than twinjet, after pioneering the introduction of the world's first wide body twins in the early 1970s with the A300/310, followed by the colossal A380 disaster.

Boeing's errors, on the contrary, stayed mostly in the narrow body and the middle of the market segments, led by stretching of the 737 program too far leading up to the 737 MAX disaster ultimately and not fielding timely & effective replacement for the 757-200 program in the middle of the market (covered later), given Boeing's laser focused strategy preoccupation of dominating the wide body aircraft market.

The Zeitgeist of commercial aviation at the start of the 21st century had been changing rapidly and had been way different from the late 20th century. The advent of next generation engine technology & a surge in air traffic driven by waves of globalization had led to the arrival of a number of new cowboys virtually riding their way into the commercial aviation crop fields as potential aircraft manufacturers, supported by their respective governments, with motivations ranging from looking to capitalize on growth opportunities (Bombardier, as the modern age's virtual aviation gunslinger always quick to the draw) to developing indigenous capabilities

& industrial base (COMAC & Irkut) to both (Mitsubishi). With them came a new herd of new, clean sheet aircraft programs targeting the regional to commercial aircraft segments but lacking the skilled & seasoned shepherds to keep & lead them on course. This exotic & eclectic mix of players included, Bombardier as the most ambitious leader of the pack, China's COMAC, long time Boeing T1 supplier & industrial partner, Mitsubishi from Japan and Russia's Irkut.

The competitive landscape of commercial aviation, thus, was on the verge of a paradigm shift which was to see the emergence of the third competitive force for a change since the disappearance of McDonnell Douglas from the sky in 1997. Most of these players, however, did not have any serious domestic aviation manufacturing capabilities on the commercial side while also lacking crucial experience of managing complex supply chains & assembly operations, with the exception of Bombardier; which was to become Boeing's nemesis in the next phase indirectly by changing the traditional rules of the game set by the very well established Airbus-Boeing duopoly as industry behemoths and big boys of the aviation landscape!

"You may delay, but time will not." – Benjamin Franklin

Chapter 7

THE FALLOUT, THREE-WAY BATTLE WITH 'REVISED HORSEPOWER' AND THE 'KRYPTONITE EFFECT'

"There is no Avoiding War; it can only be postponed to the Advantage of Others" – Machiavelli

Spurred by the Bombardier's radically revisionist attempt to change the traditional rules of the game and to disrupt the market with next generation technology and almost literally threatening the incumbents; Airbus was the first one to retaliate and to go for the counter attack, aiming to take the battle to the invader, by launching the re-engined A320neo in early December 2010 with almost 95% airframe commonality between the neo & ceo variants.

Offering an option of two engines; one being the Pratt & Whitney's PW1100G GTF engine and the other CFMI's LEAP-1A engine; the A320neo offered a 15% better fuel efficiency and 20% lower engine maintenance costs & almost a significant reduction in noise footprint at takeoff as compared to its predecessor, which was still to continue its journey for the time as the A320ceo, trudging alongside the neo down the taxi track. Airbus invested just over EUR 1 billion/USD1.3 billion[1], almost equivalent of the list price of 10 original A320ceo aircrafts, into the development of the neo version of the A320 which was scheduled to enter service in 2016 with Lufthansa as the launch customer.

The A321neo and the A319neo variants were scheduled for entry into service, following the A320neo, after precise 6 months intervals. Airbus, targeting 200 A320neo deliveries to airlines for 2017, thus, had its next blockbuster in the making, at a paltry financial investment as well as minimal risks while also having anticipated the usually expected, minor teething troubles over the initial entry into service of the new engine. Airbus had also catered for the likely demand uptick from the launch of the new, re-engined A320 family by adding further capacity in form of the new Hamburg based fourth final assembly line for the A320 family, with the new line ultimately becoming operational in July 2017.

The A320neo, however, was almost 1.8 tons heavier than its predecessor, the A320ceo, given the heavier, newer engines, minor airframe adjustments as well as the addition of sharklets (winglets) as standard which was not an issue for the neo at all which retained the same take off & landing performance with minor aerodynamic adjustments. But the complexity & gamut of potential issues for a similar, re-engined 737 were not going to be this simple, as it had been for the A320 airframe, given the fact that the A320 was two decades younger to the 737 in generational development while the 737 belonged to the 1960s aviation era. In February 2011, Boeing CEO, Jim McNerney said: "We are going to do a new airplane"[2] while Jim Albaugh, President of Boeing Commercial Airplane said a month later in March 2011 that Boeing was not really sure about re-engining the 737.

The A320neo received a grand reception at the 2011 Paris Air Show with the program booking a record 667 aircraft orders worth $60.9 billion within a week at Le Bourget[3]. With the stellar performance at the Paris Air Show, the A320neo's order book stood at 1,029 aircraft orders, within 6 months of its launch, making the A320neo the fastest selling aircraft program in the commercial aviation history, thereby, effectively putting the ball in Boeing's court while simultaneously also raising the stakes sky high for Boeing. Airbus CEO, a beaming Tom Enders commented at the Paris Air Show: "Le Bourget 2011 is a strong confirmation of our product strategy. With over 1,000 commitments just half a year after launch our A320neo is a real bestseller. I have to admit, I largely underestimated the market demand for neo before this show[3]."

Clean Sheet Vs. Re-Engining: Boeing's Dilemma

The A320neo's perfect launch & superb take-off was enough to trigger Boeing's anguish which had booked way lesser aircraft orders for the 737NG through the entire year of 2010. Boeing, having missed the opportunities available earlier to create a worthy successor for the 737, almost had its back to the wall now. Bombardier's encroachment on to the big boy's turf had altered the strategic calculus of the narrow-body segment of commercial aviation radically and the C-Series had almost given jitters to both Airbus & Boeing which had so far become accustomed to a two-person, zero-sum game play out format.

For Boeing, the available options were either to start from scratch by revisiting the drawing board for a new, cleans sheet aircraft design which could replace its 737, which was the difficult option as it would have been a $10-$15 billion bet and entailed risks of getting it right the first time besides the possibilities of a protracted development scenario, much like the 787 experience. Additionally, Boeing had still not been able to get off its exorbitant roller coaster ride on the 787 Dreamliner program, which was still gearing up for the take-off roll in early 2011 and had cost Boeing almost an astronomical sum of $32 billion in development costs (which incidentally also brought Boeing its fourth prestigious Collier Trophy award in 2011 after the 747, 757/767 & 777 programs earlier) and was potentially staring at a long & winding tail of entry into service issues which continued to haunt Boeing post the delayed 787's original entry into service in October 2011.

Having invested to the hilt to ensure 787's market success, Boeing was reluctant to plough in another $15 billion towards an entirely new aircraft program in early 2011, an option, which was further constrained by Boeing's need to play catch up with Airbus in market timing to stay on course in the narrow body race which had so far been running neck and neck and it wouldn't have been possible to keep it at least that way by going for the scratch up development option given the longer development horizon involved as against re-engining.

Boeing Plays it 'Out of Turn' and the Old 737 is Reincarnated, Yet Again!

Airbus' market strategy for the A320neo had the U.S. market, Boeing's home ground, as a key pivot and a key hunting ground and Airbus spared no effort in making deeper inroads into the U.S. market with the neo, which clearly had an overmatch over the 737NG. Airbus had its payback in full measure when American Airlines, a traditional Boeing operator, announced its plan in mid-2011 to renew its narrow body aircraft fleet with the procurement of next generation aircrafts through the decade with the move triggered by a surge in oil prices, almost a 30% spike year-on-year in 2011.

Making a radical departure from its traditional strategy of maintaining

an all Boeing fleet, American placed an inclusive, dual-source order for 460
new narrow body aircrafts worth almost $38 billion[4], rendered as one of the
largest aircraft orders in aviation history by then. The order; which also
comprised 260 aircrafts from the A320 aircraft family split equally between
130 A320ceos & 130 A320neos; almost shook the ground beneath Boeing
headquarters in Chicago, which in fact turned out to be the epicenter with
the aftershocks felt as far as Seattle!

A company press release by American Airlines announcing the aircrafts
order was actually intended at prodding Boeing. A part of the issued press
release read as below: "As part of the Boeing agreement, American will take
delivery of 100 aircraft from Boeing's current 737NG family starting in
2013, including three 737-800 options that had been exercised as of July 1,
2011. American also intends to order 100 of Boeing's expected new
evolution of the 737NG, with a new engine that would offer even more
significant fuel-efficiency gains over today's models. American is pleased to
be the first airline to commit to Boeing's new 737 family offering, which is
expected to provide a new level of economic efficiency and operational
performance, pending final confirmation of the program by Boeing. This
airplane would be powered by CFM International's LEAP-X engine[5]".

The statement, made on the occasion by Gerard Arpey, Chairman and
CEO of AMR and American Airlines, clearly was aimed at détente with
evident streaks of underlying diplomatic prowess: "This was an incredible
opportunity for our company that presented itself from two great
manufacturers. And, given our aggressive and ambitious fleet plans, we feel
fortunate to have both Boeing and Airbus standing beside us to meet our
needs[5]". This kind of an equal & inclusive approach shown publicly by a
leading U.S. legacy carrier towards its arch-rival Airbus and on top of that
coming from an all Boeing fleet operator was akin to almost a virtual

European invasion of the North American continent and was indicative of an emerging tectonic shift in market realities.

Knowing well that Airbus wouldn't stop at this and its next stop well in fact could be Southwest; another Boeing loyalist and the world's largest low-cost carrier as well as the largest operator of the 737; Boeing had to make a quick call. This was really significant given that Boeing's latest market forecast at the time, released in July 2011, forecasted a demand for 23,000 airplanes for the 2011-2031 period for the market segment in which the 737 operated, an opportunity worth a whopping $2 trillion[6].

Facing significant pressures from market forces coupled with a direct, frontal attack from a fully armed & loaded to the gills Airbus invading Boeing's very own backyard; and unwilling to concede ground & market share in an almost high stakes series decider play out; Boeing counterattacked and ultimately took the decision a month later; almost impetuously and seemingly on the fly; to go in for re-engining of the 737 with Airbus clearly having forced Boeing's hand into taking the decision indirectly. The old 737 was thus reincarnated again, for the third time in its fourth generational avatar!

Boeing announced the launch of the 737 MAX in August 2011 after getting the green light of approval from the board of directors and having received almost 496 orders from 5 airlines customers[6]. The new 737 was going to be powered by the LEAP-1B engines as the aircraft's sole power plant.

Boeing's President and CEO of the Commercial Airplanes division, Jim Albaugh, roared at the launch: "The re-engined 737 will allow Boeing to continue to deliver the most fuel efficient, most capable airplane with the lowest operating costs in the single-aisle market. This, coupled with industry leading reliability and maintainability, is what customers have told us they want. As a result, we are seeing overwhelming demand for this new and improved version of the 737. We are working with our customers to finalize these and other agreements in the weeks and months ahead."[6]

Next Round of the Battle of the Heavyweights in the Narrow Body Ring: A320neo vs. 737 MAX

For Airbus, the re-engining of the A320 was the correct call at the moment given that the airframe & core technology set; which already were a generation ahead at the program's inception; were just over 2 decades old, unlike the vintage 737; which had seen over 4 decades of service and thus its ability to sport an entirely new pair of next generation engines without getting dizzy & disoriented was dubious!

Boeing had already faced structural issues & airframe limitations while re-engining the 737 in the early 1990s at the time of developing the 737NG, the third generation of the 737, with the aircraft's low mounted wings not leaving enough ground clearance below them for the engines that were becoming larger & bigger in size with time in the relentless quest to achieve ever higher bypass ratios to become more fuel efficient.

At that time, in the 1990s scenario, CFM International had played the savior having proposed & eventually solved the issue by mounting the pair of its latest configuration CFM56-7B engines just ahead, in front of the wing rather than below the wing as is done traditionally while also reshaping the nacelle & de-rating the engine consequently. This kind of unusual position for mounting the engines on the737NG left just 18 inches of ground clearance below them (refer images below) to the ground, just barely enough for smooth take off runs.

Image 5: A Boeing 737-400's CFM56-3 engine, with its recognizable non-circular "hamster pouch" inlet;

Image Source: Davidelit, Wikimedia Commons, Public Domain

Image 6: CFM56-7B of a 737-800(NG series) – Note the proximity to the ground

Image Source: Trainler, Altair78, Wikimedia Commons, **CC BY 3.0**

The new CFM Leap 1-B engines, however, were even larger & heavier than the CFM-56-7B engines, they were to replace on the same 737 airframe. The Table 1 below captures & compares key parameters between the CFM56-7B engine (and its variants), which power the 737NG variants and the LEAP-1B, which powers the 737 MAX variants, with CFM retaining the exclusive engine supplier status on both NG & MAX.

Illustration 1: **CFM56-7B vs. LEAP-1B**

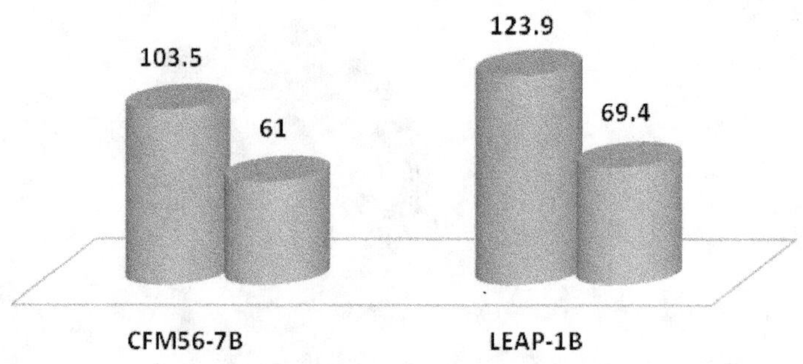

Illustration 2: **CFM56-7B vs. LEAP-1B**

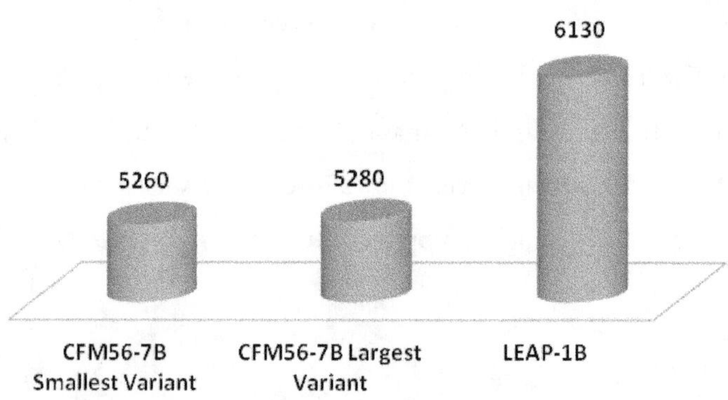

Illustration 3: CFM56-7B vs. LEAP-1B – Maximum Take-Off Thrust

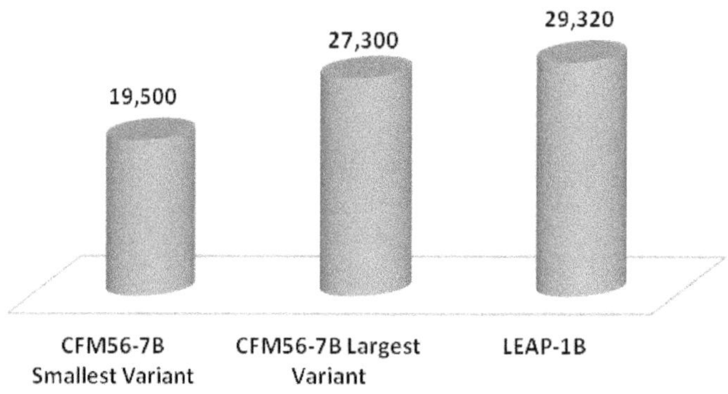

The LEAP-1B engine is almost 16% heavier than the CFM56-7B with the difference of almost 850-870 lbs/385-395 kg in dry weight between the two engines. With the 737 being a twinjet, the total difference in weight & additional weight liability on the airframe stood at around 1700-1790 lbs or around 800kg. The fan of the LEAP-1B is also 14% larger than the CFM56-7B with a difference of around 8.4 inches, a key factor in ratcheting up the bypass ratio up to 9:1 on the LEAP-1B (from 5.1 to 5.5:1 for CFM56-7B) translating into higher fuel efficiency. The LEAP-1B engine is also longer by around 20% than the CFM56-7B and produced almost 50% more thrust than the lowest variant of the CFM56-7B family and around 7.5% more than the top end variant of the CFM56-7B[16,17,18]

A longer, heavier engine with a larger fan diameter, in form of the LEAP-1B, meant significant changes & alternation in the original aerodynamic profile of the 737 airframe necessitating structural aerodynamic adjustments from the engineers. This was given the fact that the CFM International had custom-built the specifications of the LEAP-1B, in view of the 737 airframe limitations, keeping them slightly on the lower side than the other members of the LEAP engine family (see charts below).

Illustration 4: **LEAP-1A vs. 1B vs. 1C**

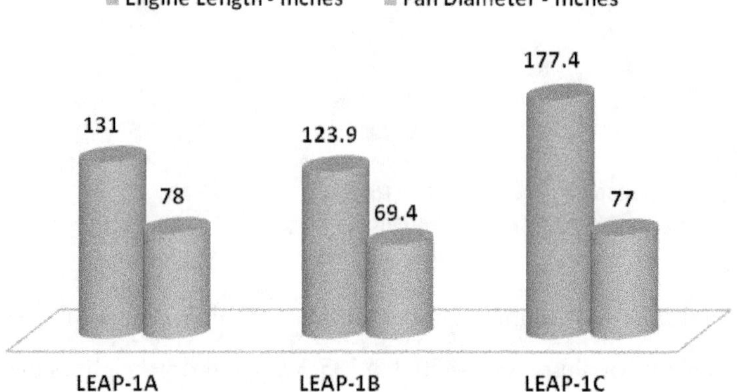

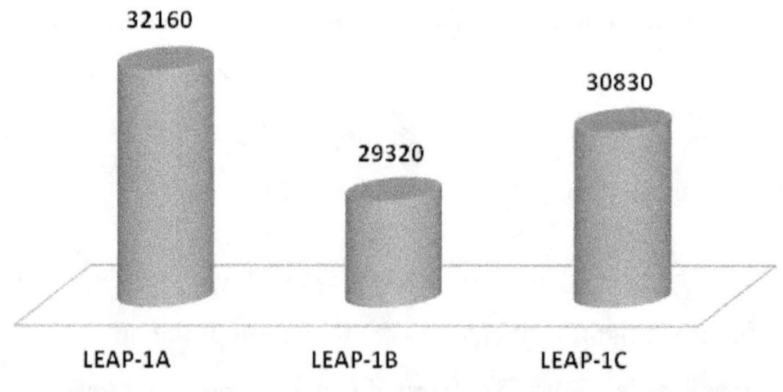

Illustration 5: **LEAP-1A vs. 1B vs. 1C**

LEAP - 1A v.s 1B vs. 1C Variants

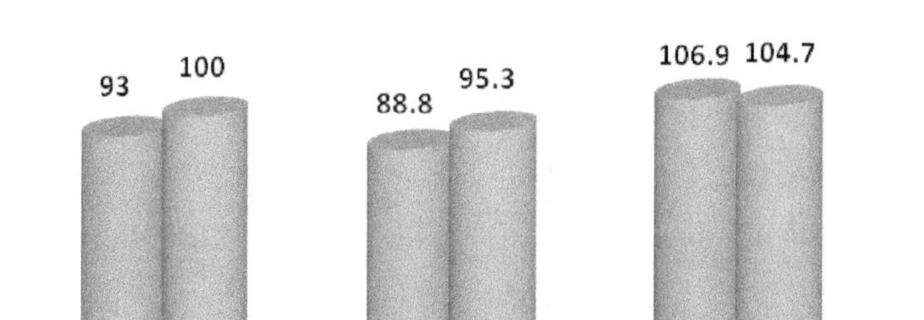

Illustration 6: **LEAP-1A vs. 1B vs. 1C**

It is important to look at this comparison amongst the LEAP engine family variants here for two reasons. First, the comparative view indirectly underscores the structural limitations of the over 40 years old 737 airframe as gauged from the engine specifications developed by CFMI for it which were almost at the lowest end of the spectrum spanning the triad of LEAP engines family. The LEAP-1B had the lowest, sub-optimized bypass ratio of 9:1, as against 11:1 for the 1A & 1C variants, translating into a slight difference in fuel efficiency if not offset by other mechanisms in the engine architecture, given the size & structural modifications tailor-made for the 737 airframe.

Second, the LEAP-1A engine variant powers the 737's arch rival, the A320neo, as an engine option while the LEAP-1C variant powers the brand new, under development narrow body flying unicorn from the orient, the COMAC's C919, a potential future competitor to both the A320neo & the 737. The point to see here is the huge dimensions of the -1C engine built for the C919, which are much larger, by a factor of almost 20%-30%[17,18], than the -1A & -1B built for the A320neo & the 737 MAX respectively.

The mounting of the CFM56-7B engines on the 737 airframe directly underneath the wings in their usual position was not possible even in the 1990s, thus, ruling out any possibility of the larger LEAP-1Bs being mounted there at all. The last known good configuration, therefore, was chosen but the engines had to be mounted even further ahead on the wings & slightly higher relatively, than the 737NG, given the ground clearance issue.

However, the almost 800kg of additional weight penalty brought in by the new LEAP-1B engines along with their larger dimensions & higher thrust output brought in abrupt changes to the airframe's original aerodynamic profile by shifting the aircraft's original centre of gravity & its handling characteristics by inducing an aerodynamic effect leading to pitching up of the aircraft's nose upwards during steep climb scenarios translating into the airplane's tendency of stalling mid-air with a sudden reduction in lift to the airplane wings leading to potential aircraft control issues.

The issue had been pinned down to the structurally larger and higher thrust generating new LEAP-1B engines which induced this tendency along with their unusual mounting position on the wings[7,8] (those more technically aligned, may refer to the referred source links for a deeper dive into how the interplay of thrust force & direction of force created this effect). This must have been evident to Boeing's engineers during the wind tunnel testing of the scale model itself even before progressing to the prototyping stage.

Image 7: Mounting position of CFM56 Engine on a Boeing 737NG (below)

Image Source: Aviacsa, Wikimedia Commons, Public Domain (PD)

Image 8: Mounting position of LEAP-1B on a Norwegian Air
International's Boeing 737 MAX (below)

Image Source: Edward Russell, Wikimedia Commons, C-C-A 2.0 Generic
License

Since the problem was limited to specific, steep climb situations involving high angles of attack; a structural, hardware based aerodynamic solution may not have been the obvious choice as its incorporation would have further altered the aerodynamic profile of the aircraft.

Boeing, running out of time and patience, chose to borrow a leaf out of the McDonnell Douglas' classic aviation playbook and went with the active deployment of cutting corners approach, led by the secret insertion of the Maneuvering Characteristics Augmentation System (MCAS) software piece; with Boeing terming it as a flight control law; into the MAX controls with absolute authority (discovered by the pilots & airlines painfully much later) to manage & counteract the effect, creating its own version of the DC-10 in the process for the twenty first century aviation with Boeing's management ringmasters overruling its engineering base outright.

The MCAS system, upon detecting a high angle of attack for the aircraft based on Angle of Attack (AoA) sensors, initiates adjustment to the horizontal stabilizer which in turn pushes the aircraft's nose down to correct the high angle of attack as an automated corrective measure. This MCAS action on MAX was configured to be based on data emanating from a Single Angle of Attack sensor (of the two on-board the 737 MAX) rather than the usual two or three sensors on aircrafts leaving it susceptible to erroneous readings & a potential catastrophe in the making as a single point of failure.

Further, Boeing enabled this mechanism to operate with full authority & repetitively in almost a looping pattern while also enabling its activation even at low airspeeds with the removal of G-Force variable from the activation triggers. The MCAS on the 737 MAX once activated was meant to pull the aircraft's nose down for 10 seconds at a stretch followed by a 5 seconds gap and then another 10 seconds of nose down bid in almost a looping pattern till the aircraft & those onboard succumbed.

This was to the extent that it could virtually 'hijack' the airplane without the pilots being able to override the system as they were not even aware of the presence of this kind of a Frankenstein system on-board let alone being capable of tackling & handling it. This was going to drive the pilots absolutely crazy & leave them clueless about the aircraft's sudden & abrupt seizure of control from them.

The antecedents of the MCAS could be traced back to the KC-46A Pegasus aerial tanker program; which had been under development in parallel at Boeing's Everett, Washington based facilities (Everett) as the 737 MAX (Renton) during the same phase; with the KC-46A program won by Boeing based on its 767 program derivative. Boeing had used the MCAS for the first time on the KC-46A to counter some aerodynamic issues & improve handling characteristics. However, on the KC-46A, the MCAS derives data from twin sensors (instead of one on MAX) & is enabled with full override mechanisms by the pilots and lastly as well as most importantly, the USAF pilots are aware of its onboard presence and have been trained to manage it effectively. Seemingly, this is how the seed, containing the idea of installing MCAS on MAX as a quick fix, originated ab initio, germinated quickly and was almost obsessively followed up ultimately by an utter botched up implementation. An article appearing in the New York Times has covered the background story on the MCAS; it's poorly planned & executed insertion onto the 737 MAX and the overall functioning in great depth.

The MCAS problem was further compounded on MAX by the fact that Boeing made vehement efforts to keep the same type certificate on the 737 MAX, as the earlier generations of the 737 aircraft series, through the certification process for the MAX adopted by the Federal Aviation Administration (FAA). A common type certificate for airworthiness; issued by the National Aviation Authority of the State, to which the aircraft manufacturer belongs; is awarded to aircrafts belonging to a particular, same aircraft category based on aircraft design, like the A320 aircraft family, featuring significant commonality of structure, design, systems etc.

This common type certificate signifies that the aircrafts are similar to one another in operations and thus obviates the need for an entirely separate certification process altogether from scratch entailing a much longer & comprehensive process of testing, evaluations, assessments & procedural framework.

A common type certificate for MAX would have meant no additional classroom or simulator based training for the pilots considered & treated by some airlines as a hurdle as well as avoidable costs with their bottom line focused operating structures as well as the elaborate SOPs.

Boeing feared that this additional pilot training cost scenario & factor coupled with the potentially longer certification & entry into service horizon would be a significant disadvantage and would undermine the competitiveness & commercial prospects of the MAX against the arch-rival

A320neo (which was going to surely have the common type certificate and had already been on a sales rampage) in the hyper competitive airlines market and enough to be capable of shifting the odds & the overall narrow body segment's strategic calculus in favor of Airbus going forward.

These fears were further exacerbated & corroborated by the A320neo's robust & burgeoning order book position (1,000 new aircraft orders worth $60 billion by June 2011 within 6 months of program launch while Boeing was still at Ground Zero) which had been experiencing a phenomenal growth rate.

Some industry factions[19] have estimated these additional pilot training costs for the MAX in the range of $1,000 to $2000 per head as a ballpark figure which would have translated into; taking the pilot strength of 9,700 for Southwest Airlines (as per Southwest Airlines Pilots Association or SWAPA); an all 737 carrier, a still manageable amount of $9.7 million to $19.4 million at the most for Southwest alone. However, the mere prospect of a staunch Boeing loyalist, Southwest, switching to Airbus camp would surely have been Boeing's worst nightmare.

Boeing, thus, in its efforts to keep the same type certificate, hid the addition of the MCAS system on the 737 MAX from the regulators as well as the pilots who were to fly them. The addition of the MCAS to the 737 MAX was a significant departure from the traditional 737 series of aircrafts all of which do not have 'Fly by Wire' (FBW) controls meaning that the pilots instructions are executed through a mechanical way comprising of pulleys & cables, unlike the FBW, where these instructions are executed by the computers onboard the aircraft.

The 737, thus, traditionally had been a pilot's aircraft where they could call all the shots as they wanted and fly it based on the feel. The addition of the MCAS on the MAX, however, took the 737 series' flight controls from a traditional manual system into the somewhat of a semi autonomous terrain which would literally have spooked the pilots flying them later and definitely necessitating awareness of the presence of such a system on-board to tackle it coupled with additional training hours to be able to execute & replicate the steps to tackle it during actual passenger flights within the required response timeframe.

However, Boeing; determined to maintain the same type rating in order to stick around in its narrow body catch up game with Airbus and to effectively bypass the need for fresh certification & the associated pilot training costs preposterously did not even mention the term 'MCAS' even once in the aircraft's technical literature sent to regulators or the aircraft flight manual.

Boeing instead created a two and a half hours short computer based training program for existing 737 pilots, which could be undertaken by them all by themselves even just prior to flying, all in all a sham in the name of pilot training & on-boarding on to the new aircraft ultimately leaving the pilots as well as the passengers unabashedly at the mercy of a half baked software system based solution!

The 737 MAX, saddled with the burden of Boeing's evidently care-a-damn attitude towards safety, marked by a series of process lapses & goof ups and circumventing its way through a shoddy certification process made its way to the skies receiving FAA certification in March 2017; based on the original type certificate issued for the archaic 737-100 & 737-200 aircrafts over 4 decades back in 1967; with the first 737 MAX delivery taking place two months later in May 2017 to Indonesia's Lion Air subsidiary, Malindo Air, which was yet to pay the price for it in full measure!

What an infra dig for the "Spirit of Renton"…

The Kryptonite Effect

This kind of a shoddy, reckless & outrageous approach to airplane development & modification must surely have drawn the ire of the legends of the American aviation industry of the yore, even in heavens up above, especially, iconic American airplane designers, including, the Wright Brothers themselves, Howard Hughes, Kelly Johnson & Boeing's very own airplane designer duo, Jack Steiner and Joe Sutter; all of whom literally shaped the industry's evolution radically with their trailblazing engineering minds, apart from intrepid American explorer aviators like Wiley Post, Will Rogers & Amelia Earhart who simply dedicated their lives to aviation.

The origin of the emergence of this kind of third-rate approach to engineering, however, could be traced back to Boeing's mega acquisition of the McDonnell Douglas, the original masters of the 'Cutting Corners Engineering' philosophy. McDonnell Douglas had this culture of 'Numbers First' which put financial profits over everything else, reflected amply in the safety track record of their airplanes over decades of commercial aviation service.

The pack was spearheaded by the notorious wide body, DC-10 trijet, which developed a highly undesirable & infamous reputation as a highly dangerous airplane over its poor safety record subsequent to involvement in multiple, horrendous air crashes through the 1970s; with the tally of DC-10 crashes standing at 61 through its service lifespan with the last one being as recent as October 2016[2]; owing to multiple design flaws which were subsequently identified & rectified over the decades that followed.

The 'Mad Dog', MD-80, the loud & noisy single aisle airliner powered by a pair of fuselage mounted Pratt & Whitney JT8D turbofans had been a narrow body workhorse for the airlines that operated it. However, it too was highly prone to developing rudder blanking risks leading to potential loss of directional control upon application of reverse thrusters right after touchdown apart from reporting of multiple incidents of jackscrew assembly troubles and malfunctioning of landing gear.

Some of the MD-80 variants; led by the MD88s and MD90s; had still been in-service with some U.S. based carriers till COVID-19 happened. The ones operated by Delta Airlines (a MD loyalist), witnessed their farewell flights in mid-2020 while American flew them till September 2019 and they were in fact cheered with a broad range of mixed emotions & love hate reactions with some pilots & crews getting overwhelmed while some other factions terming it as good riddance.

Folks at McDonnell Douglas were masters of shorts cuts, quick fixes, stop gap measures & reusability of designs as well as parts that went into their airplanes and the following statements & opinions shared by multiple airlines executives go on to further corroborate that view. "Going back to the DC-9, even parts of the DC-3 made their way into it (referring to the MD-88 & 90). Some of the knobs go all the way back to the 1930s. Douglas was really good about using things over & over. So much aviation history was handed down" said Kurt Tahara, a first officer with Delta for over 4 years[10]. Sam Mayer, a pilot, who was promoted to Captain on the MD-80 in 1999, at American Airlines said: "I remember thinking it was the most un-ergonomic cockpit I had ever seen, with switches and dials seemingly placed at random spots all over the cockpit, bearing no semblance to any other aircraft I had ever seen[11]". "It's almost as if they built the airplane and, at the very end, finished up and said, 'We forgot to put the compass in.' It's a relatively goofy system, but it worked."[11] said Kent Wien, an American Airlines captain who used to fly the MD-80, referring to the highly odd placement & positioning of the compass behind the co-pilot's seat in the MD-80 to avoid interference from other cockpit equipment. This was obvious & along the expected lines given the operating model of McDonnell Douglas for commercial side of business was akin to the 'Flying Circus' guys who would hire engineers just to set up the show and would later fire them once it was up & running.

The acquisition of McDonnell Douglas by Boeing in 1997 brought in a number of exotic & wild ponies, in the form of stalwart defense programs into Boeing's more or less civilized stable at the time with most of them packing & adding a lot of punch to its defense portfolio; the likes of AH-64 Apache, F-15 Strike Eagle and the F/A-18 E/F Super Hornet; which are still in production & service to this day.

However, it also brought in the unseen kryptonite ore in inheritance along with the kryptonians which started permeating and impacting generational crop of engineers & managers inside Boeing from thereon undetected & unchecked by Boeing leadership with no internal, organizational firewalls in place. This kryptonite was the McDonnell Douglas' 'Numbers First' culture which put financial profits over everything else, including, safety, which was in direct contrast with Boeing's aboriginal organizational culture of 'Engineering & Safety First'.

The people & personnel, or the 'Kryptonians', who had landed on Boeing's deck from McDonnell Douglas had this culture deeply ingrained in their DNAs. They interacted with Boeing's vastly different, 'engineering focused' culture, maneuvering their way through to seize & control core power & nerve centers within Boeing. This charge was commanded by the Ex-CEO of McDonnell Douglas, Harry C. Stonecipher, who became Boeing's President and COO, along with Michael Sears, who had served as McDonnell Douglas' head of Commercial Aircraft and Defense & Space businesses earlier, and became Boeing's CFO without having any formal background in finance. This numbers focused culture somewhat gradually started eclipsing & supplanting Boeing's original engineering culture with a culture in which profits & shareholder returns reigned supreme across the board.

Stan Sorscher, a physicist & workers union negotiator who has worked for Boeing for more than two decades said: "A long & proud 'safety culture' was being replaced with 'a culture of financial bullshit, a culture of groupthink"[12] in an interview in 2019.

Further, Boeing's technical side of the brain, its engineers & technicians, had staged a major strike in the year 2000 with them protesting against the prevailing work culture at Boeing and looking for improvements demanding more involvement & a say in decision-making as engineers felt that their ideas were falling on deaf ear & being turned down[13] with Boeing's focus solely on cost cutting following its acquisition of MD, spearheaded by Stonecipher, who had adopted the same tactics & modus operandi earlier at McDonnell Douglas as well. A union leader protesting during the strike even said to press later: "We weren't fighting against Boeing, We were fighting to save Boeing[14]." The statements above clearly reflect the impact the infusion of McDonnell Douglas culture had started having inside Boeing.

The negotiations around all other demands were accepted and the deal was deemed as success but the battle for the culture, however, was lost in the process as the return of the striking workers caused fissures amongst the workforce at large and led to emergence of trust issues with the management precipitating into attrition issues, ultimately calling for firefighting measures from Boeing which provided transient first-aid measures which proved to be good only for the surface bruises and only for the time being. Boeing's decision to outsource most of the design & production work on the 787 program in early 2000s added more fuel to this fire as the 'technical side of the brain' inside Boeing was already feeling frustrated & jaded with Boeing having built only one new clean sheet airplane, the 777, over the past two decades since 1985, till the 787 was announced in 2004.

In September 2008, over 27,000 Boeing machinists staged a major and one of the longest strikes (since 1995) over issues with outsourcing, job security and pay & benefits as a direct fall out of the 787 outsourcing model & decisions. The strike lasted for almost 8 weeks and cost Boeing almost $100 million[15] per day in revenue terms coupled with a $7 million hit to profits on a daily basis, including, charges booked on account of delays in aircraft deliveries.

Boeing's aircraft deliveries for 2008 dropped by 15% year on year, after a record breaking performance in 2007, with Boeing's stock price also receiving a major battering. This drove the incumbent Boeing Chairman & CEO, James McNerney, to establish Boeing's first industrial presence on the U.S. east coast in form of the second 787 final assembly line (FAL) in a non-unionized North Charleston, South Carolina in 2010; Boeing's first FAL outside the Washington state; to counter the negotiating heft of the workers' unions with the pickets virtually turning into a permanent & almost invisible as well as insurmountable picket fence between the warring sides.

A statement by Peter Morton, a retired Boeing Vice President who started out as an engineer on the 737 and was the marketing manager on the 737 program from 1969-1974, given in a media interview in 2016 underscores & captures the essence of these persistent engineering vs. management power struggles within Boeing effectively. He said, "Boeing's a complicated company. You can have a team of engineers improving the airplane and at the same time you have the bean-counters running the numbers on closing the program"[17].

Seemingly, that familiar & longstanding 'Andon Cord' between the Boeing's management echelons and its engineering base had been snapping and almost beyond repair and it was going to have significant ramifications going forward. The situation by 2013 had been no different with Boeing reverting to its cost cutting rhetoric post the 787's $32 billion financial mess and haggling hard over workforce wages despite strong revenues & bottom line growth trends with Commercial Aviation super cruising in one of its longest demand upswing cycle.

An old chestnut, which had been circulating widely in Seattle at the time of the Boeing-McDonnell Douglas merger, perfectly captured the essence of the deal quintessentially as "McDonnell Douglas bought Boeing with Boeing's money.[14]" From here on, it was just a matter of time before it ultimately was to become Boeing's Achilles' heel & bring the giant's moment of fall from grace!

The addition of elements of McDonnell Douglas' company logo, post modifications, to Boeing's original company logo post acquisition; showcasing an aircraft encircling the globe paying tribute to a Douglas aircraft's first aerial circumnavigation of the earth in 1924; had been the harbinger of what was to come, the sign of the stormy clouds headed Boeing's way..

These engineering vs. management power squabbles & wrangling over issues, sometimes also pertaining to matters coming purely & exclusively under the purview of the former's jurisdiction within large organizations have not been a very rare phenomenon.

Another great institution which has faced this typical tug of war recurrently on its decks in the past has been the U.S. Navy. Strikingly similar face-offs between the Navy's over- empowered line officers and the engineers & technicians base; over design, technical & other engineering specific issues and their subsequent outcomes for the Navy translating directly into capability gaps & shortfalls; have been elaborated brilliantly, eloquently and in great depth by the Late Admiral Hyman G. Rickover; a great persona as well as a towering figure, the 'Father of the Nuclear Navy' and someone whose career has been underscored by an unparalleled safety record while operating a nuclear fleet; in one of his speeches, 'The Role of Engineering in Navy',[16] given in 1974. A read is highly & definitely recommended for those inclined for some invaluable lessons & pearls of wisdom from 'The Kindly Old Gentleman'!

"Better to have one thousand enemies outside the house than to have one single enemy inside it". – Lebanese Proverb

Chapter 8

A 'MESSED-UP' DECADE AND A SHEEPDOG HERDING A 'PACK OF WOLVES'

"Even the Finest Sword plunged into Salt Water will eventually Rust". – Sun Tzu, The Art of War

The years in transition at the turn of the century were tumultuous for Boeing with the company booking financial write-offs amounting to billions of dollars without board's timely & effective intervention or course correction measures while making abrupt & radical strategy moves spearheaded by an overarching focus on making the shift towards achieving portfolio balance through diversification & further expansion in the defense side of the business. This was happening in a decade characterized by a historic defense spending meltdown in a post Cold war world. Last and the least wanted highlight of the period was Boeing's involvement in multiple scandals and the questionable ethical conduct of its executives, especially, the two Chairman & CEOs, who had reigned through this nightmarish decade for Boeing, namely, Philip Condit and Harry C. Stonecipher.

Strategy Shift Geared towards Portfolio Balance

Starting with the strategy piece, Boeing gave a major fillip to its defense business in the 1990s through multiple acquisitions, worth almost $20 billion cumulatively, with the most notable ones being the Rockwell's Aerospace & Defense business (worth $3.1 billion) in 1996, followed by a defense heavy McDonnell Douglas in 1997 (worth $13.3 billion) and Hughes' space & communications businesses in 2000 worth $3.5 billion.

Boeing's this deep dive into the defense business was a major departure from its traditional forte of commercial aircraft manufacturing and was a bit incongruous with the prevailing market context. The reason being that the 1990s was one of the worst decades for defense spending given that the Cold War was formally over with the defense industrial base tottering & almost in dire straits marked by an unprecedented industry consolidation. The defense asset base had literally been on fire sale with industry consolidation in full swing and driving the M&A activity up through the roof which added Martin to Lockheed, Grumman to Northrop and brought Rockwell & McDonnell Douglas on to Boeing's deck.

The deals struck by Boeing were structurally sound with the portfolios of Rockwell Aerospace & McDonnell Douglas being a good strategic fit for Boeing clearly with Rockwell's & McDonnell Douglas' defense heavy portfolios complementing Boeing's commercial heavy portfolio perfectly along with the existence of significant synergies growth potential that could be harnessed by combining the commercial airplane business portfolios of both Boeing & McDonnell Douglas (with their combined total order backlog of around $100 billion[4]). Additionally, Boeing was getting the elusive option of obliterating its key & long time U.S. based commercial airplane business rival for decades, McDonnell Douglas, by going through with the nation's tenth largest merger (a $13.3 billion all stock deal) in size at the time.

This was accompanied by the opportunity to become the world's largest integrated commercial & defense aerospace behemoth in the process despite a contraction led near term downswing view of the defense spending trajectory appearing clearly & evidently on the screen. Boeing almost took a long shot here amid prevailing low visibility levels with the sun having almost already set on the defense industrial base in the Western Hemisphere towards the end of the 20th century till a new radical & asymmetric threat emerged on the horizon at the very onset of the 21st century.

However, the rationale behind a sudden & radical strategy shift, led by defense-focused acquisitions, was driven by Boeing's broadening of vision combined with the potential ability of a balanced portfolio to be able to provide a clear & effective hedge and offset the highly cyclical nature of commercial aviation rather than any outright apparent growth avenues visible on the radar in form of any big ticket defense programs with the exception of the F-35 JSF.

Boeing had just unveiled its 20 year strategy vision in 1996, spanning the 1996-2016 horizon, with its incoming (in 1996) CEO Philip Condit outlining Boeing's vision & plans to become an integrated aerospace player going forward, underpinned by the theme of 'Connect & Protect[2]', covering the entire gamut & spectrum ranging from commercial to military aviation to space systems. The strategic focus clearly had been on generating sustainable high returns & value for shareholders rather than staying limited to and being content with riding the market tides to generate merely market cycle driven returns.

A comment by Jerry King, Boeing's President of Defense & Space Group in 1996, on Rockwell acquisition further underscored & echoed the command objectives. He said: "It accelerates us to achieving our 20-year vision, which calls for Boeing to be a fully integrated aerospace company designing, producing and supporting commercial airplanes, defense systems and defense & civil space systems. We anticipate that our defense and space businesses will continue to grow[3]". The potential growth opportunities mentioned by King in his comment referred to "the Joint Strike Fighter program, Airborne Lasers and GPS Satellites[3]". Of these, Boeing ultimately lost the diamond mine, the Joint Strike Fighter program to Lockheed Martin within the next 5 years, lasers are still under development (in the year 2020) even after 2 decades and the last satellite production contract Boeing received from the USAF for GPS IIF satellites was in 1996 and for the GPS III satellites Boeing chose not to bid in 2018 with the GPS III, too, ultimately going Lockheed Martin's way!

Amongst the military aircraft programs coming from the McDonnell Douglas stable, the F-15 Eagle & F/A-18 Hornet, were 4th generation fighter jet platforms which were past their prime and were slated to be replaced by the Joint Strike Fighter program, the variants of which were to be procured en masse for services. The potentially lucrative MRO cash flow streams likely to originate down the road from the large, in-service fleets of F-15s and the F/A-18s over the following decades were Boeing's key focus areas in military aviation with the F/A-18 Super Hornet program just on the verge of entering service with the U.S. Navy in 1999.

However, delays on the F-35 program have provided the 'Eagles' & the 'Hornets' with the much needed mid-air refueling, thereby, extending their ability to remain airborne for much longer besides keeping their production lines humming. They are still & very much an integral component of the respective service's force structure & fighting capabilities, contrary to Boeing's CEO Condit's conservative expectations in 2001, when he said that he expected the F-18 to remain in production at least 'through the end of that decade[2]'. Boeing's latest win with the USAF's F-15EX contract in July 2020 provides a fresh lease of life to one of the USAF's best air superiority fighters with an impeccable track record. Military rotary beasts, especially, Hughes built attack helicopter AH-64 Apache & Vertol's CH-47

Chinook, too, have equally been top of the line and still remain highly lethal
warriors & potent cogs in the U.S. & allied war machines post upgrades and
still ruling international exports in terms of market shares in their respective
segments. The C-17 Globemaster III again was top notch in the military
transport aircraft segment as the strategic lifter of choice in the western
hemisphere, however, it's assembly line was the first one of all to be wound
up in 2015 with the order book drying up.

Boeing, thus, had rapidly built up a well-scaled & developed defense
portfolio with alternate industry cyclicality in less than a decade to
effectively shield itself from the commercial aviation's swinging, roller
coaster ride with the aggressive pursuit of an inorganic growth strategy
executed full throttle.

By 2001, Boeing had more than doubled its top line this way from
'$25.4 billion for 1993 to $58.2 billion for 2001'[2] with the contribution of
commercial airplane segment to Boeing's top line dropping drastically from
'80% in 1993 to 60% by 2001'[2] , which had came down further to 50% by
2010 further reducing the portfolio concentration risks significantly.

Boeing's defense & space business, thus, quadruped in size from a
revenue base of $5.4 billion for 1993 to around $25 billion by the year
2001[2], led by the aggressive pursuit of M&A strategy. This came after a
rocky start to the decade of 1990s for Boeing's defense & space business;
which had accumulated operating losses of $1 billion towards the end of the
year 1991[4], prompting the in-command CEO, Frank Schrontz, to go in the
overdrive mode with his chainsaw on a major cost cutting spree to right
size the segment in-line with the market realities.

The onset of a pronounced downswing in commercial aviation post the
9/11 attacks & the concurrent uptick in defense activity in 2001
immediately made the strategic rationale behind portfolio diversification

self-evident in action. In the post 9/11 aftershocks, Boeing's commercial airplane segment's top line plummeted from the peak of $35 billion for 2001 to $22 billion by 2005 while the defense business zoomed up from a revenue base of $22.8 billion for 2001 to $30.8 billion for 2005, cushioning the same effectively. However, some of the apples hand-picked by Condit were 'B grade' as well. For instance, the Hughes space & communications, which Boeing had bought for $3.75 billion, had major production & quality issues and the commercial satellite market did not turn out as he had expected resulting in Boeing booking a financial write down of $1.1 billion.

Shifts in Culture as well as 'Home Ground'

The other element of the strategy piece was the marked shift in Boeing's culture from 'the slow & ponderous mindset led program focused culture' to the 'relentlessly & rigorously business focused culture with focus on speed & agility' as outlined in the Chairman & CEO Philip Condit and Vice Chairman Harry Stonecipher's combined 'Message to Shareholders', in Boeing's 2001 Annual Report.

This included a strategic maneuver executed at supersonic speed with Boeing's world headquarters moved in a jiffy from Seattle to Chicago (a city which didn't even have traces of Aerospace manufacturing) from conception to completion within a 167 days span (between March 2001 & September 2001) leveraging a hefty free cash flow generation by Boeing amounting to $2.7 billion for 2001[2].

That included another structural organizational change with the business heads of Boeing's core businesses moved up & elevated to the position of CEOs of those segments in an apparent shift towards decentralization with greater authority to be vested with the Tactical Leadership for the perceived speed & agility gains.

This strategy of moving the Corporate Centre far away from the
company's all major production hubs & nerve centers; located at Seattle in
Washington, the hub of Boeing's commercial airplanes business production,
Long Beach in California which produced the C-17 Globemaster III and St.
Louis in Missouri, the seat of Boeing's defense, space & security business;
which spearheaded the day to day operations and were the very theatres
where the ranks & files were actually deployed, was unfathomable and
intended actually at isolating them in turn. The reigning Chairman and the
Vice Chairman collectively unraveled its underpinnings to be the proposed
shift from day to day, operational & tactical focus towards the overarching
vision & strategy with 'no meddling in operational stuff' as the rationale
ostensibly.

The critics & industry observers, however, have termed it as a deliberate,
well planned stratagem by the 'Kryptonians' in the lead who were looking
to rapidly push & dislodge Boeing away from its traditional 'Engineering
Focus' towards the 'Profit Focus' orientation; which had been ingrained &
coded deeply into their DNAs, courtesy the most illustrious & precise
leadership production machinery of the era, General Electric (GE); with
Stonecipher being the prime example. This was being attempted at a
lightning pace; aimed at practically turning & reorienting the chess board
their way with a classic execution & display of management legerdemain
while the onlookers were at bay. Unfortunately for Boeing, they succeeded
in the coup attempt, courtesy weak leadership at the helm and a comatose
board on-board!

Stonecipher also created the Boeing's very own, sprawling, centralized
leadership centre based in St. Louis, Missouri at a facility he himself had
bought in 1995 for McDonnell Douglas, as its CEO. The centre was in-turn
a replica of the GE's model-learning center based in Crotonville, New York

and its original, quantitative focused 'Sledgehammer & Anvil' leadership
model. The Crotonville centre; with its genesis & strategic positioning as a
key organizational change lever and a legacy dating back to the 1950s; had
been leveraged by GE's stalwart CEO, Jack Welch, as an effective
organizational change fulcrum in the 1980s to catalyze, fundamentally alter
& reboot GE's traditional leadership genetic code by producing the next
generational production batches of radically re-engineered & re-oriented
leaders equipped with a genetically modified leadership DNA focused on
shareholder value at the core and a brand new bushido issued for reference
containing & outlining best practices for navigating a globalized business
landscape effectively.

The idea behind leveraging the centre for inducing a radical,
organization-wide change was to institutionalize the practices Welsh had
himself pioneered & pursued while carrying out radical transformation of
GE in the 1980s spearheaded by a major downsizing covering almost
100,000 employees, divestments worth $6 billion and acquisitions worth
$13 billion in a span of a mere half a decade in the 1980s (1981-1986)[5]
ultimately giving a sort of turbo boost to GE's market value. An article
appearing in 1989, titled 'GE's Crotonville: A Staging Ground for
Corporate Revolution'[5] by Noel M. Tichy; who had himself led the
Crotonville centre through 1985-87; throws a spotlight on the 'leadership
revolution & stagecraft tutelage'[5] under way at Crotonville at the time.

The best part of the whole thing was that the Boeing leadership centre
was also being run by ex-GE guys! The centre was set up in March 1999 at
an investment of $60 million with Stonecipher delivering the centre's
opening leadership sermon himself as the Boeing's COO[6]. Stonecipher,
upon his re-ingress as the Boeing CEO later in 2003, as a replacement for
Condit, himself said: "When people say I changed the culture of Boeing;

that was the intent, so that it's run like a business rather than a great engineering firm[7]".

However, the relevance & contextual appropriateness of GE's leadership model; built originally & configured for GE's huge conglomerate structure & model; for the niche, predominantly engineering focused culture of a typical commercial airplane manufacturer at the time, like Boeing, is the million dollar question for the Organization Development domain & its practitioners. Further, its rapid introduction at almost hypersonic speeds into an organization right after implementation of major organizational changes (defense acquisition spree that brought in traces of vastly different organizational cultures) without a well laid out cultural integration strategy coupled with the physical as well as mental hijacking of the Corporate Centre; all of which collectively triggered tectonic internal shifts, the aftershocks & reverberations of which continue to impact Boeing even after almost over two decades now!

Leadership Whittle

Philip Condit was an avid aviation enthusiast and an engineer with a strong technical bent at the core of his persona having earned the pilot's license at the age of 18 and filing a key patent on a flexible wing design in 1965 at the outset of his career with Boeing as an aerodynamics engineer. Starting from the grassroots as a fresh engineer holding a Master's in aeronautical engineering from Princeton, he carved his way up through the ranks with his exceptional engineering acumen with significant design leadership on the 757 & 777 programs ultimately going on to become Boeing's 7th Chairman in 1997.

However, he was always the engineer and aviation geek at heart, always enamored by airplane design & technology breakthroughs despite summiting the highest echelons of leadership at Boeing. His psychographic

silhouette & profile sketch, as drawn up by a cover story appearing in
Bloomberg Week[8] at his abrupt exit from Boeing in 2003, describes him as
'frequently indecisive, isolated, aloof & non-confrontational[8]' in a stark
contrast to & very unlike Boeing's tradition of strong & commanding
leaders.

This also was in complete contrast with Stonecipher, the second in
command at Boeing, who was suave, outspoken and 'decisive & brutally
candid[8]' which set up the stage perfectly for the rise of McDonnell Douglas
culture and its ultimate supplanting of Boeing's own Engineering culture. In
fact, Condit actually might have been a better cut to lead an advanced,
specialized research or technology development division within Boeing, like
the Phantom Works (which came later with McDonnell Douglas) instead of
leading a wolf pack of acquired C-Suite executives!

The Counter-Attack Strategy that Backfired and the Mega 'Financial Write Downs'

Apart from the personal issues, Condit also made the strategy faux pas
in his first year as Chairman, of boosting production rates abruptly &
significantly for commercial airplanes at Boeing's Seattle based plants. The
move was part of an aggressive counterattack strategy pursued by Condit to
take on a surging Airbus in the North American market, as per which,
Boeing was giving aggressive discounts to customers and had been filling
up the order book rapidly while trying to make up for the low average
profit per aircraft by reducing manufacturing costs by almost 25% while
doubling up on production rates to offset the same. The plan to ratchet up
production rates at lower overall costs of production had been hatched up
to counter Airbus' tactic of under cutting Boeing on pricing and had in turn
led to a boom in Boeing's order intake.

The strategy had much deeper roots which could almost be traced back to the early 1990s, to the tutelage of Boeing's CEO Frank Shrontz's tenure4. Shrontz; fearing Airbus' growing heft in North America during the 1990s with a pricing led market strategy; fired the warning shots and made cutting production costs by around 25% a key imperative to maintain competitiveness to take on a spirited Airbus which had been selling airplanes at huge discounts in its bid to make deeper market inroads in North America.

He also was the brain behind the preposterous idea of signing exclusive 20 year procurement agreements for Boeing airplanes with the then leading four U.S. carriers in 1993 to check Airbus maneuvers, which nevertheless were pursued, fructified & honored by them as well!

Boeing also had been giving accelerated airplane delivery slots to airlines apart from discounts in a demand upswing cycle and by the end of 1997 Boeing had a robust order backlog of $94 billion[9] for commercial airplanes with Boeing having booked new orders for 5029 commercial airplanes in 1997. To meet the aggressive delivery schedules the production rate on the 737 program had been ramped up dramatically and more than doubled from 20 aircrafts a month in early 1997 to almost 45 aircrafts a month within a year's span by early 1998 with Condit planning to ramp it up further to 47 a month from Q2 1998[9].

Boeing's supply chain, however, could not cope up with this frantic pace and in turn buckled down & succumbed to the pressure and as a result the line fell flat & was cold for almost 30 days with parts shortfalls & production disruptions impacting the 737 & 747 programs severely and resulting in a financial write-off of $2.6 billion announced by Boeing in late 199710, marking the first major setback for Boeing in the 20th century in its raging war against Airbus for aerial supremacy.

This was followed by another write down of $1.4 billion announced by Boeing in January 199810 towards the remaining aircraft inventory acquired as part of the McDonnell Douglas merger. These assets had to be written off with Boeing deciding to pull the plug on the MD-80 & MD-90 aircraft programs. This led to Boeing booking its first & the largest financial loss in almost 40 years since 1959, amounting to $500 million for Q3 1997 & $178 million for the year 1997 which was in stark contrast with Boeing's net income of $1.6 billion for the year 1996[10]. With the debacle, Boeing lost almost $4 billion in market value while also wrecking its North American aerospace suppliers in the process.

Adding the financial write downs on the Hughes commercial satellite business to the above, the cumulative tally of financial write-offs booked by Boeing during Condit's tenure turned out to be almost $5 billion. Add to this the simultaneous development of the re-engined 737NG & clean sheet 777 by Boeing in the early to mid-1990s with the 777 program, launched by Boeing in October 1990 with an initial development budget of $6 billion, ending up overshooting it by a large margin. By the late 1990s, Boeing had already scrambled its top generals and tasked them with the mission to develop a radically different war doctrine to win aircraft development wars quickly and in a cost effective manner going forward, which in turn, ultimately ended up leading to the genesis of a financial disaster, in form of the 787, in the next decade!

The Ethical Meltdown and the Ensuing Implosion

The last nail in the coffin for Philip Condit's career came in form of the scandals which rocked Boeing during the opening years of the 21st century and awakened the board as well which had been operating in a virtual comatose state so far. The first scandal; involving procedural irregularities & ethical violations in the procurement plan for 767 aerial tankers by the

USAF; led to conviction & prison sentence for Boeing CFO Michael Sears and Vice President Darleen Druyun (who had joined Boeing in January 2003 after leaving the USAF) in October 2004. Boeing also had to pay millions in financial penalties over the scandal.

The second major scandal, which rocked Boeing's ship in 2003, triggered launch of even more fireballs, regulatory action & further reputational damage for Boeing. Some Boeing engineers, who had originally joined from Lockheed Martin, were found to be in possession of proprietary data & information pertaining to Lockheed Martin's space launch program running in thousands of pages which could have given an unfair advantage to Boeing for the U.S. Air Force's Evolved Expendable Launch Vehicle (EELV) Program, initiated in 1995 and potentially worth $15 billion over 20 years.[11]

Boeing had been competing head-on (via McDonnell Douglas post acquisition) with Lockheed Martin over the EELV program. Deeper investigations led to the Department of Defense ousting Boeing from key space launches worth $1 billion[11] while also barring its rocket division, acquired originally from Rocketdyne in 1996, from participating in future government contracts.

The board, which had so far accepted a series of multi-billion dollars worth of write-offs, too, had been awakened from its deep slumber by the shockwaves which followed the public blasting of Boeing by the media as well as the public at large over the twin scandals. Phil Condit took moral responsibility of all that had transpired, albeit with none of the proceedings indicting him, and in Boeing's interest, he put in his papers and disembarked the ship after firing CFO Michael Sears, who was later incarcerated. Stonecipher's return to the scene was no saving grace for Boeing either and in 2005 he also had to be ousted over ethical issues over his involvement & affair with a fellow Boeing employee. In fact, Condit's overall track record in this personal aspect of his life was no better than Stonecipher!

Boeing, thus, had two of its CEOs exiting and a convicted CFO serving prison term behind bars over ethical issues within a span of 2 years, i.e. from 2003-2005 and this was nothing short of an utter cultural & ethical meltdown as well as implosion of the highest intensity for one of the few great organizations of the yore still left standing tall as the true beacon of American industrial might. What a rocky start to the 21st century for Boeing!

Condit's legacy

Condit could be best credited for adding might to Boeing's portfolio with the rapid scaling & beefing up of the defense side of business and the subsequent portfolio broadening & balancing it brought in apart from elevating the top line rapidly albeit with the cultural integration challenges & issues it brought along inadvertently. However, he is also accused by critics of letting Airbus, which had been nibbling at the heels for market share, to go scot free, maneuvering unmarked & unchecked on the field to be able to push up the score-line in its favor rapidly and ultimately going on to build its aviation empire in the 1990s decade; eventually becoming an equal rival with almost a 50% market share by the turn of the century, which had risen sharply up from 30%, while Boeing was busy with its extensive shopping spree down the defense high street!

He is also believed to have misread Airbus' wide-body strategy in the late 1990s, led by the A380 program, by firmly believing that Airbus would never launch the A380 program while paying heed to absurd proposals, which included the Sonic Cruiser and a further stretched variant of 747, the 747-X, towards the late 1990s for a potential counteroffensive. His successor, Stonecipher's multiple, radical decision choices in the crucial launch phase of the 787 program effectively laid the keel for the $32 billion colossal financial mess which Boeing is still dealing with even to this day.

All in all, it was a messed up decade for Boeing which could at best be forgotten lest for the expensive & precious lessons it entailed for the next waves of leaders to come onboard the vaunted ship.

Condit's legacy: A balanced portfolio with the beefing up of defense business and the cultural erosion & its ultimate mutation following a coup!

Rise of Airbus as a Worthy and Equal Rival

Airbus had its catapult phase in the mid-1990s when it really took off following the first major expansion of its product portfolio with the broadening of wide body aircraft line-up with the addition & entry into service of the A340 in 1993 followed by the A330 in 1994 with Airbus, thereby, adding new flying horses to its stable which had been occupied so far by the A300/310 workhorses.

On the narrow body side of things, the first & the largest sibling, the A321, had joined the A320, thereby, creating the A320 family in 1994 with a common type certificate followed by the addition of A319 in 1996. The Airbus sales machinery, too, had gained critical mass by the mid-1990s following a decade of successful market disruption, especially in the U.S. market, led by a maverick John Leahy who subsequently applied the model & his playbook to the global aviation chessboard.

Airbus' aircraft deliveries[14] for the first half of the 1990s decade, 1990-

1994, stood at 676 aircrafts going rapidly up from the around 100 aircrafts level annually for 1989 & 1990 to crossing the 150 aircraft deliveries threshold for the years 1991 & 1992 (163 & 157 aircraft units respectively). Of these, the A320 program which had just entered service a year back in 1988, accounted for over 70% (73% and 71% for 1991 & 1992 with 119 & 111 aircraft deliveries and almost doubling from 58 aircraft deliveries for 1988 & 1989). This strong uptick in aircraft deliveries was preceded by a massive surge in order intake at Airbus which booked cumulative orders for 825 aircrafts worth $61 billion collectively for the years 1989 &1990[14]. Airbus, in fact, booked gross orders for 421 & 404 aircrafts in 1989 & 1990 worth almost $34 billion & $27 billion respectively and it had a lot to do with the A320 program which just had its EIS in 1988 and enabled Airbus to effectively penetrate the market on which Boeing had its stranglehold and make deeper inroads as also unraveled by the Airbus' overall aircraft deliveries mix for the 1990s decade below. Also, propelled by the A320, Airbus expanded its operator base rapidly by more than two fold from a mere 68 operators for 1990 to 178 operators by 1999[14].

Some of the key, strategic deals cracked by Airbus for the A320 family for the 1990s decade, with some of them being really smoking deals structured to undercut Boeing on pricing & financing terms, included the $3.2 billion deal[16] with Northwest for 100 A320s which though was signed originally in 1986 but was confirmed ultimately at the 1990's Farnborough Airshow. This was followed by the landmark & breakthrough lease deal[17] signed with United in 1992 for 50 aircrafts with options for another 50 worth $2.4 billion marking & rounding off Airbus' successful penetration of the U.S. market with most leading U.S. carriers having come into the Airbus fold by the early 1990s, which included, American, Continental, Delta, TWA, Northwest, United and America West.

This was enabled by the fact that A320 had been strategically positioned & very effectively targeted by Airbus at the 727-200 trijet, DC-9 & MD-80 twinjet replacement market in the U.S. hinged on the A320's superb fuel efficiency; courtesy the high-bypass CFM56-5B engines; which was almost double that of the low-bypass JT8D powered in-service fleet of 727-200 and the DC-9/MD-80 apart from the additional seating capacity, much lower noise levels and enhanced operating range the A320 offered in direct, head-to-head comparisons. The 737 Classic series led by 737-400, too, was an under-match for the A320 with its relatively much limited operating range of 2,060 nmi as against the A320's 3,300 nmi, which, post the United deal, forced Boeing to sound the bugle for the genesis of the comprehensively revamped & re-engined 737NG series, which was ultimately launched in 1993.

The key limitations of the 737-400 in effectively taking on a generationally advanced A320 in range & capacity, which must have been obvious to Boeing at the A320 launch in 1984 itself, should have been countered with a proactive response from Boeing with either a 737NG decision in the mid-1980s itself or a fine-tuned 7J7 configured precisely for that role which could have effectively thwarted the Airbus' coup in the narrow body segment. However, Boeing's reactivity in this phase of the game in the second half of the 1980s gave Airbus just enough leeway and tactical maneuvering space it needed on the runway to take-off and never look back again!

The icing on the cake for Airbus came in August 1998 in form of the crackerjack of a deal[15], the largest in the European consortium's history (by that time), for 188 A320s powered by V2500s worth $9 billion[15] at list prices signed with another staunch Boeing customer, British Airways, which placed its first aircraft orders with Airbus, to replace its existing 737-200s, in

a deal sweetened & clinched with huge discounts. The victory for Airbus,
however, came after a bitter dogfight with Boeing which had aggressively
pitched its latest 737NG series which had just been certified and entered
service.

Airbus also reached the key milestone of first 1,000 cumulative aircraft
deliveries (since inception) in 1993, which had grown rapidly by almost 50%
within a 3 year span, up from 652 (cumulative deliveries) by the end of
1990, with air traffic receiving a major boost following the formal end of
Cold War and the onset of a massive wave of globalization.

The second half of the 1990s decade, i.e. 1994-1999, proved to be even
more propitious for Airbus with aircraft deliveries surging by almost 40%
to 955 aircrafts[14] cumulatively as compared to 1990-1994, wherein Airbus
had delivered 676 aircrafts[14]. Also, Airbus' annual aircraft deliveries peaked
in 1999 and almost touched the 300 level (with 294[19] aircrafts delivered in
1999). Thus, within the span of a decade, Airbus' annual aircraft deliveries
had tripled from the 100 aircrafts level for 1990 to almost 300 by 1999.

Additionally, Airbus booked total gross orders for 804[19] aircrafts worth
almost $59 billion[14] for the years 1989 & 1990 with 50% of the orders (for
446[4] aircrafts[19]) belonging to the A320 family alone (A320 & A321
programs). For the second half of the decade, 1995-1999, Airbus doubled
that tally by booking orders for 1,924[19] aircrafts worth almost $130 billion[14]
with the A320 family accounting for an almost 80% share with 1525[19]
aircraft orders.

Airbus delivered a total of 1631[19] aircrafts for the 1990s decade (1990-
1999)[14], of which the A320 family collectively accounted for almost 65%

---

[4] These included 61 A320 orders which had been previously received by Airbus.

with around 1000 aircraft deliveries (1068)[14],[19] led by the A320 program, which alone had a 44% share of the total aircraft deliveries with 718[19] aircrafts! Compare the 1068[19] A320 family aircraft deliveries made by Airbus to the 1782[18] 737 aircrafts delivered by Boeing for the 1990s decade and the rise of a heavyweight in the making could be easily figured out. The A320 program thus effectively & unequivocally became the much needed springboard in Airbus' incredible rise from a rookie to become an equal rival in the global commercial aircraft market within a span of almost 3 decades by the turn of the century.

The McNerney's Decade: GE-derived Leadership DNA, Competition Focused Strategy and a Shareholders Focused Game Plan

Boeing somehow could never get rid of the traces of the GE-derived leadership DNA which was induced into Boeing with the acquisition of McDonnell Douglas in 1997 and in turn caused mutation of Boeing's original cultural DNA. The mega merger, orchestrated by Harry C. Stonecipher himself, effectively brought him into Boeing's cockpit as the first officer joining a well entrenched but shaky Philip Condit in the Captain's seat. Stonecipher had been heading the GE's aircraft engines business in the mid-1980s and then became the President and CEO of Sundstrand & later the Chairman as well. Post the exit of both Condit & Stonecipher over the twin scandals; James McNerney Jr., who had already been serving on Boeing's board since 2001, was brought in as the Boeing's new Chairman and CEO in June 2005.

However, that did not make much of a difference in either leadership style or approach with the 'GE Way' continuing to influence things inside Boeing. Prior to joining Boeing, McNerney had been the 3M's Chairman and CEO and in the past he had also been the CEO of GE's aircraft engines business during the 1980s, having been through Jack Welch's ideological indoctrination & leadership tutelage at Crotonville. In fact, McNerney had been one of top three stallions competing for Jack Welch's position upon his retirement with Jeffrey Immelt ultimately getting the top job. The GE pedigree & legacy thus had effectively been carried forward nonetheless. Additionally, McNerney had been a Harvard graduate but was the only Boeing Chairman to not have a formal educational background in Aviation or engineering for that matter.

The first half of McNerney's decade-long reign was caught up in shepherding a troubled 787 program back on course and getting it into commercial revenue service somehow and post that he had markedly expressed his disdain for the traditional approach; spanning over a decade and worth multi-billion dollars; to new aircraft development programs with his famous 'moon shots' remark.

With the commercial aviation in a marked upswing cycle driven by a low crude oil price environment & strong passenger traffic growth; McNerney chalked out his game plan for the next decade pivoted on ramping up production rates across aircraft programs. The plan was to boost up production output & aircraft deliveries significantly while lowering the overall cost base leveraging scale advantages effectively to boost revenues as well as profitability. In fact, overall aircraft production rates at Boeing actually rocketed up by almost 50% by 2015 as against the 2010 level.

The side B of the plan was to reduce procurement costs further by putting Boeing's negotiating heft to full effect and keeping suppliers on their toes while playing hardball with the trade & workers unions as well while signing long term contracts with them. The final piece was to prioritize & focus on low-cost & low-risk derivative aircraft programs rather than clean sheet programs with reference to overall product strategy with the ultimately objective being to keep shareholders happy with large dividend payouts and continued share price bolstering through stock repurchases.

The decision to go for re-engining the 737NG into MAX rather than taking a clean sheet approach in 2011 was somehow guided to a large extent by McNerney's this clear repugnance for clean sheet approach to aircraft development post the 787 ordeal given the fact that Boeing had squandered billions in sunk costs which were going to impact program's breakeven & eventual profitability. Further, his non-engineering background, need to address the call of the hour to retaliate & lock horns with the A320neo at the earliest to stay in the battle as an active combatant apart from the perfect alignment of the derivative strategy approach with his shareholders focused operating philosophy collectively influenced & ultimately shaped the MAX decision.

The other two re-engined aircraft programs launched under his reign included the 747-8 (launched in November 2005) following the 787's launch in 2004 and the 777X (launched in November 2013) right after the 737 MAX launch in August 2011. The clean sheet NMA plan too had been brought under contemplation at almost the fag end of his tenure which ended in 2015. Given that the 737 MAX has been a disaster, 787 a financial mess, 747-8 & the KC-46A being in the forward losses zone, NMA being a no show and the 777X delayed & under development with a stagnant order book in a turbulent aviation market; one tends to question the achievements & legacy of the McNerney's decade from the product strategy perspective, especially in the light of his huge pay package, millions doled

out to him in annual bonuses and showering of stock awards by Boeing
followed by provisions for substantial post retirement pension benefits.

However, the achievements of the McNerney's decade & his overall
game plan might be found in another place and that is Boeing's financial
statements which might provide more clues to the real achievements made
during his reign. During the first half of his tenure, Boeing's commercial
airplane segment's revenue base grew by almost 40% from $22.6 billion for
2005 to $31.8 billion for 2010 while Boeing's top line grew by just under
20% from $55 billion for 2005 to $64 billion by 2010.

By the last full year of his tenure, which was 2014, Boeing's top line had
reached $90.7 billion growing up by a whopping 50% from $64 billion for
2010, led clearly by the Commercial Airplanes segment, whose revenues
had doubled from $31.8 billion for 2010 to $60 billion for 2014 while the
BDS top line had remained more or less static. Boeing's asset base, too, had
grown by around 50% from $68 billion to almost $100 billion by 2014.
Boeing's share price zoomed up by almost 83% during his reign from
around $70 at the end of 2005 to around $128 by the end of 2014 along
with market capitalization.

All of this had been achieved by a relatively simple game plan, the rapid
ramp-up of aircraft production rates from a mere, just-under 300 aircrafts
level for the years 2003-2005 (290 for 2005) to the just below 500 units level
by 2010 (462 for 2010) going further up to 723 aircrafts by 2014. The
production rates at Boeing, thus, increased almost two-and-a-half fold
between 2005 & 2014 while they grew further up by almost 50% between
2010 and 2014.

The plan had been undergirded by the fact that commercial aviation had begun its oscillatory motion & journey towards the full upswing cycle end; post the onset of recovery from the 9/11 aftermath starting in 2006 followed by the deep dive off the cliff in 2009 triggered by the 2008 global financial crisis; driven by forces of global trade & globalization and thus it was time for Boeing to put the pedal to the metal. Boeing's 20 year market forecast released in 2010 pegged the long term average growth rate for passenger & cargo traffic at 5% with Boeing projecting a demand for 30,900 new airplanes worth $3.6 trillion over 20 years, as per Boeing's Annual Report for 2010.

Another key point is the split of these aircraft delivery numbers by programs with the 737 evidently & clearly leading from the front in volume terms. The 737 program accounted for almost 73% of Boeing's total aircraft deliveries (in units) for 2005. For the year 2010, the 737 program's share stood at 80% of the aircraft deliveries while for 2014 the contribution stood at around 67%.

Interestingly, a majority of the profits & wealth generated went straight to the shareholders by way of fat dividend payouts, bonuses, bonus shares and stock price appreciation led by an aggressive pursuit of share buybacks duly authorized by the Boeing's board. In October 2007, Boeing's board authorized common stock repurchases amounting to $7 billion. In December 2014, the board authorized another mega stock repurchase plan worth $12 billion while a Category 5 Tropical Storm (which was to eventually turn into a Super Typhoon), named 'MAX,' was already building up around the Seattle area. In fact, astonishingly, Boeing's cumulative outlay in returns towards the shareholders over 2013-2019 (during McNerney & his successor Muilenburg's tenures collectively) stood at a whopping $60 billion[13] in dividend payouts and share repurchases!

McNerney's game plan might have been deemed & termed a huge success from the numbers perspective and by the wild exuberance of the animal spirits driving & ruling the stock markets which made him thrive at the throne effectively for almost a decade. However, his long term product strategy choices could be termed at best as sub-optimal euphemistically and paved the way for the current mess Boeing has ended up creating for itself with the top performing stallion, 737, perfectly rewarded with a tawdry & slapdash generational Avatar!

McNerney's innings brings another point to the fore or rather a dichotomy or a trade-off between the strategy choices in front of a typical CEO of either adopting a short term strategic posture focused on quarterly earnings as well as near term shareholder returns and stock price movement to keep the ever so impatient bulls & the sadistic bears happy by simply milking the company's cash cows & feeding them or taking a long term strategy approach focused on the company's future and positing it on the road towards long term success by taking tough calls, marshalling troops, leading from the front in combat and going above & beyond the call of duty while enduring outcomes of difficult decisions which might be painful financially or otherwise over near term but always in the long term interest of the organization.

However, in the world & age of activist investors; with their growing appetite, clout & tendencies of meddling in to almost dictating the companies' strategies & affairs with a bought seat on the board; it has been more of a utopia for the business world and better left for the armed forces where honor still holds sway above the materialistic avarice. That's why some CEOs across industries who have tried to walk this 'path of fire' have ended up being fired eventually, trampled in turn by the stampede of raging bulls forever seeking perennial bull-runs and crushed by the sadistic bear-hugs!

However, there have been stark exceptions to the generalization just made about the business world with the same epitomized & exemplified best by none other than the business tycoon, Chuck Feeney; the Co-Founder of the Duty Free Shoppers Group in the 1960s; who has quietly donated his entire wealth, amounting to almost $8 billion, to charities over the past four decades with his exceptional 'Giving while Living' philosophy inspiring & getting adopted by the business moguls, Bill Gates & Warren Buffet, who are actively continuing to pave & tread this path least travelled. However, the other exception about Feeney is that he, too, is a war veteran at the core having served the USAF with a short stint in the Korean War!

Coming to the military side, even the armed forces have not been an exception either with the 'Father of the Nuclear Navy', the highly decorated Late Admiral H.G. Rickover, having to scream at the top of his voice perennially (even after his departure post 63 long & distinguished years of service) to warn the Navy of the impending perils of letting a culture dominated by line officers & their battery of outsourced contractors' self-serving approaches & stratagems prevail over the Navy's traditional, inexorable pursuit of technical excellence culture for ensuring long term success with a sustained overmatch over enemies.

The outcomes for the Navy have been more than self-evident in the 21st century with the Littoral Combat Ship (LCS) program being just the 'tip of the iceberg'. The LCS has been an embattled program plagued by multiple design & operational issues encountered within days or weeks of entry into service apart from battling serious capability shortfalls, cost overruns & contractors led maintenance issues with almost over $30 billion already down the drain & no future pathway in sight.

The other big & worthy name which could be added to the list has been the Gerald R. Ford-class aircraft carrier program and the USS Gerald Ford, CVN-78; the most expensive aircraft carrier ever built at $18 billion with a 24% cost overrun than originally budgeted and delayed by years over a plethora of issues, most noticeably with its nuclear propulsion system, and still struggling & far from entering full, active service with the Navy.

Further, serious command & training issues on the Navy's decks have been raised & highlighted by investigation reports over the twin deadly collisions involving Navy destroyers in 2017; the USS Fitzgerald destroyer (that collided with the container ship MV ACX Crystal in June 2017 off the coast of Japan) and the USS John S. McCain near Singapore two months later in August of the same year; with both fatal accidents collectively leaving 17 sailors dead. The deep probes launched post accidents have found major leadership, decision-making and crew training & preparedness deficiencies on the Navy's part.

The GE promulgated leadership model; skewed highly towards the shareholders & investors terrain; has perfect alignment with the modern age business structure & its financial underpinnings and that perfectly goes on to explain the success & the widespread acclaim it has enjoyed in the form of action & replication over decades since Jack Welch pioneered & perfected it in the 1980s.

However, somehow it prioritizes near term returns, by keeping them above everything else, and sometimes even brutally trades as well as predates the long term view & pathway for it. The prevailing state of affairs at GE over the past couple of years have gone on to show that no one is infallible & invincible with the very same shareholders; whom it has fed with choicest of the financial savories, caviar & multi-course meals to capacity over decades; after its blood for the dismantling of the conglomerate ultimately into multiple companies over divisional underperformances to unlock shareholder value and having already overthrown Jeffrey Immelt in a 2017 investor coup.

This followed the ousting & firing of his successor John L. Flannery by the board a year later in 2018 with GE stock, operating in the nosedive mode, reaching its nadir towards the end of 2018 having already fallen almost by 50% over 2016-2018 followed by staging of a recovery in 2019 and GE continuing to operate in course correction & consolidation mode. Every model has its merits as well as demerits and it is in fact just a matter of time & change of overall strategic & operating context that the lopsidedness of any perspective becomes self evident!

McNerney's Legacy: Simply, the Game Plan and the Product Strategy Mess!

The Airbus Deck – An Aerial View

The Airbus (then EADS) deck, in the 1990s decade, was also no Napoleonic era's elite & venerated 'Garde Impériale' regiments either from a relative perspective. What it was could have been best described as a mixed bag of deployed regiments of the regular French Army and its elite 'Légion étrangère' or the 'Foreign Legion' comprising of & teeming with a diverse & multi-lingual infantry derived from parts of Europe & outside the continent as well and characterized by the ensuing typical & classic heterogeneity issues marked by factionalism, structural fissures & power struggles.

The French Army irons them out through rigorous trainings but Airbus and EU didn't have the kind of jurisprudence & absolute jurisdiction as available to the Army. Airbus leadership command & decision making processes were characteristic of a typical jousting contest between the French and the German side of the leadership, fully armed & strategically positioned with their Leclerc & Panzer formations right across the fence facing each other and never shying of a potential face-off.

The bitter tug-of-war battles & constant seesaws for one-upmanship played amongst partner nations over work share agreements in the 1980s; epitomized by the wrangling seen over the decision for establishment of the A320 program's second assembly line for A321 production at Hamburg in the late 1980s with the French bickering over inefficiencies involved and the Germans campaigning actively for it citing long term gains; have been classic highlights of a politically gamified Airbus deck during the initial decades.

Airbus had the joint leadership model at the time with the French & the German CEO counterparts firmly positioned in the cockpit as the riled co-pilots ever looking over the shoulder to prevent jilting by the other while trying to chart the course & then steer it somehow. While grappling for a greater share of power & control, Airbus leadership also made strategic errors of its own kind in the 1990s decade starting with the decision to mount four engines on the A340 wide body, spurred by their relentless wild hunt chase of the 747, while the market was moving towards the twinjet direction led by ETOPS regulatory changes.

This was followed by the decision to go in for the huge flying cruise liner, the A380, again to eclipse & wrest the 747's glory with its coveted & proclaimed 'Queen of the Skies' title amid the rapidly turning market tide

which they clearly missed. The build-up of the defense side of the business
for Airbus (then EADS) started with the creation of Eurocopter in 1992
through the merger of helicopter businesses of Aérospatiale and
DaimlerChrysler Aerospace (DASA). The major push towards
consolidation of European industrial base, however, came only in the late
1990s with the dawning of the reality of an oversupply led industry scenario
amid a shrinking defense budgetary environment that drove multiple
defense side M&As geared towards consolidation.

EADS, at the start of the 21st century in 2001, still was just almost half
in size as compared to Boeing with EADS' top line at EUR 30.8 billion for
2001 as compared to Boeing's $58.2 billion at the time post the supersonic
$20 billion acquisition spree of the late 1990s. The defense business of
EADS was almost EUR 6 billion in size while Boeing's was around four
times larger at around $25 billion in 2001.

A proposed mega merger in 2012 between European behemoths, a
Civil heavy EADS & Defense heavy BAE Systems, worth almost EUR 35
Billion; could have turned the entire equation on its head and would have
created the world's largest integrated aerospace & defense company with a
potential revenue base of around $90 billion in 2012, which would almost
have been way larger than the Boeing's $68 billion top line for 2011.

This could have further complicated things for Boeing given a pivotal
role played by BAE Systems in the Boeing's global supply chain and its
sworn arch-rivalry with Airbus. Fortunately for Boeing, the deal ultimately
got torpedoed by the excellent marksmanship of a skeptical & precise
German leadership over their discomfort with the French ownership in the
merged entity and the location of the headquarters post the proposed
merger, which they wanted it to be Munich.

This political factionalism & persistent tussles inside the organization based on nationality became the fountainhead of inefficiencies & impacted ultimate outcomes within Airbus. The re-wiring issue on the A380 program (which made a huge financial dent of billions to Airbus bottom line and beset the A380 program by delaying it by over 2 years) was the classic highlight & effectively underscored the potential perils & pitfalls of this unique governance model which in many ways was based on the classic European 'Balance of Power' concept and replicated & mirrored the structural foundations, scaffoldings & underpinnings of the European Union integration model as well to a large extent along with its inherent structural limitations.

However, the issues on the Airbus deck by then were comparatively much less explosive & noisy in nature with the same holding true for their environmental & eco-system footprint as well as impact than Boeing and operated more quietly & subtly deep underneath the surface with any ripples rarely appearing onto the surface and barely visible to the naked eye!

The biggest public issue to have rocked the Airbus/EADS ship so far was the delaying of the A380 program over re-wiring issue in 2006. The A380 teams working across Toulose & Hamburg were using different versions of the product design software which led to issues and almost 500km of wiring on the A380 had to be redone as Airbus had simultaneously been working in parallel on aircraft development, customizations for customers and ramping up of production.

This announcement of developmental delays & rescheduled deliveries opened a can of worms causing a huge uproar followed by fire & fury resulting in significant capital value eradication of EADS stock on capital markets triggering a noisy feud with BAE Systems which was on the verge of offloading its stake in EADS at the time and felt targeted. This was followed by compensatory payments to airline customers, amounting to millions, by EADS along with serious charges of insider trading leveled against the incumbent CEO, Noël Forgeard, over his exercise of company's stock options worth millions just prior to the public announcement of A380 delays leading to opening of regulatory probes. Forgeard had been appointed as EADS' first President and CEO in 2001, coming onboard from the Lagardere Group, which had been the EADS' largest shareholder by 2006.

Other repercussions & ramifications of the issue included the development of a serious rift with BAE Systems and a class action suit filed by a herd of French institutional shareholders over company's handling of information to shareholders on the potential financial implications of the rescheduling of the A380 program & delay in aircraft deliveries. This raging inferno paved the way for EADS CEO Noël Forgeard and Airbus (the commercial aircraft business segment) CEO Gustav Humbert's departure with both of them stepping down ultimately in early July 2006 and clearing the way for the then French National Railways, SNCF, Chief Louis Gallois' ingress at the helm.

The credit of bridging & surmounting these factionalism issues & structural fissures, that had impaired EADS' functioning & strategic responsiveness as an organization in the 20th century, through a sweeping overhaul and galvanizing Airbus into a streamlined, fine tuned organizational entity operating as a well oiled machine & a formidable force in aerospace; free of perennial political meddling in decision-making that it had become accustomed to; goes to its former CEO, Tom Enders, to a large extent.

He had served as the Co-CEO, first along with Noël Forgeard & later with Louis Gallois, but later renounced this position and when EADS, as a fall out of the 2006 crisis, initiated a major overhaul under the hood starting with its corporate governance structure in 2007; which ultimately led to the jettisoning of the antiquated & absurd Double CEO & Double Chairmanship structure (created along the lines of the Classic European political lynchpin, the 'Balance of Power' concept); he was appointed as the CEO of the Airbus commercial airliner producing arm.

Airbus' overall game plan for the 2005-2015 decade had been much more balanced, in a typical European way, and was less aggressive in pursuit as compared to Boeing. However, it too had aircraft deliveries ramp up at the core of it with Airbus witnessing a 35% increase in aircraft deliveries over 2005-2010 from 378 for 2005 to 510 by 2010 with the same growing by another 25% to reach 635 by 2015 (as against 50% for Boeing for 2010-2014 ending at 762 by the end of 2014). This was also because Airbus had its resources fully engaged on the development of two clean sheet aircraft programs during 2005-2010, namely, the A380 and the A350XWB while Boeing had been sculpting out the 787 with a global & eclectic team of rookie craftsmen. Airbus, thus, ultimately achieved a 68% increase in aircraft deliveries over the 2005-2015 decade while Boeing clocked a 149% increase for 2005-2014!

Airbus, thus, managed to do well on the product strategy front (with the exception of the A380) while ramping up production rates in a complex industrial base set-up distributed across nations & continents; despite the continued, internal power plays & game of thrones; as highlighted by the ability to somewhat proactively (relative to Boeing) steer & achieve the generational evolution of the A320 successfully in 2010 with the A320neo,

successful creation, entry into service & market reception of the A350XWB
with Airbus delivering the goods while working in a pressure cooker
followed by a brilliant comeback with waves of lethal, full frontal assaults
on the middle of the market spearheaded by the A321LR & the XLR
through the rest of the decade.

However, power plays & struggles still prevailed even later and
persisted throughout the next decade (2010s) but they became internal
instead of being waged across the fence; most noticeably the constant tug
of war & squabbles between the German origin CEO, Tom Enders, who
had been the company CEO since 2012 and the French COO, Fabrice
Bregier; which received adequate media coverage & have been chronicled
eloquently by the press over the years. Also, EADS ultimately renamed
itself as Airbus effectively from 2014 with the company going with its best
known marquee brand with a strong global connect for an umbrella
branding strategy aimed at better cohesion & integration and with a
simplified business structure with only three business divisions, namely,
commercial aircrafts, defence & space and helicopters.

2018 was a defining year for Airbus which saw a major change in chain
of command and passing of the leadership baton aboard Airbus deck to the
next generation with the simultaneous departure of Bregier and the
retirement of the super sales chief John Leahy followed by the retirement of
the CEO Enders a year later in April 2019. In the meanwhile, there have
also been charges of corruption & violations of export controls leveled
against Airbus over multiple commercial & defense deals entailing payouts
to intermediaries for sales deals which followed the initiation & launch of
multiple probe salvos by France, Britain and the U.S. against the company
in 2016. Airbus negotiated a clear passage for itself & agreed to settle the

same with a EUR 3 billion settlement announced in early 2020[12] after having already zeroed down upon & having fired around a hundred employees post its own internal investigation into the allegations & charges. Airbus has in fact booked a charge of EUR 3.6 Billion in its books for the year 2019 towards settlement. Another probe by the German authorities, however, is still open.

"The difficulty of tactical maneuvering consists in turning the devious into the direct and misfortune into gain" –Sun Tzu, The Art of War

Chapter – 9

THE 'PRANCING MUSTANG' LANDS-UP IN THE 'FRENCH ALPS' AND 'SETTLES DOWN'

'When you surround an Army, leave an outlet free. Don not Press a desperate foe too hard". – Sun Tzu, Art of War.

The third key industry force in commercial aviation; appearing for the first time in the twenty first century with the foray of Canada's Bombardier Aerospace, positioned at the lower end of the single aisle market with its ambitious C-Series program; had gotten off to a decent start and had been shaping up well so far despite brief trysts with spurts of turbulence and had the big boys of the town all spooked up..

Unable to deal with the combined force field of the Zeigarnik and Ovsiankina effects; which had been bringing Bombardier back to the C-Series program over & over again; Bombardier launched the C-Series program finally at the Farnborough International Airshow in July 2008.

This was followed by the announcement of the selection of key suppliers for the C-Series program and the locations of industrial facilities where the development & manufacturing activities were to be carried out.

Bombardier announced Mirabel as the location for the final assembly of the C-Series while wing development & production was going to be done at Belfast. The production of aft fuselage & cockpit was to take place at Saint-Laurent, Quebec. Pratt & Whitney Canada was named as the sole source supplier of the new power plant for the C-Series which was to assemble the PW1500G engine at its Mirabel, Quebec based facility. AVIC's affiliate Shenyang Aircraft Corporation was to supply the fuselage.

Other key suppliers named by Bombardier for the program included: Honeywell for APU, Rockwell Collins for avionics, Liebherr for landing gear, Zodiac Aerospace for interiors, UTC Aerospace for electrical & actuation systems, Goodrich Corporation for engine nacelles & Meggitt for wheels & brakes.

In November 2009, Bombardier's reassessment pegged the program development cost for the C-Series program at $3.5 billion, which had already gone up by almost two third, compared against the $2.1 billion originally assessed in 2005.

Bombardier outlined the program schedule with the maiden flight for CS100 slated for late 2012 to be followed by aircraft deliveries from 2013 with the larger CS300's deliveries scheduled to start in 2014.

This ambitious program schedule was maintained by Bombardier till almost October 2012. However, in November 2012, Bombardier announced a delay of 6 months to the CS100 schedule with the maiden flight rescheduled for June 2013 and entry into service of the CS100 subsequently rescheduled for 2014. The reason for delay was described by the company as originating from issues at the unspecified suppliers end.

In the meanwhile, Bombardier also had been developing the Belfast site, which was going to be the hub for producing the wings for the C-Series program.

Bombardier announced a $860 million towards expansion & development of the new Belfast facility in November 2009[1]. During the development stage of the program, there had been multiple instances of reporting by media sources about the potential contemplation & development by Bombardier of a stretched version of the C-Series variants with extra seating capacity & operating range.

For instance, JP Morgan reported in January 2010 that Bombardier had been considering a 150 seat version of the C-Series. Bombardier, however, had rubbished the piece terming it as purely speculative. This was followed by the talks of the development of a 160 seat configuration CS300 by Bombardier for Air Asia with the talks between Bombardier and Air Asia for the same at the Farnborough Airshow 2012 reported by media[2].

Media also reported the addition of the configuration specs to the CS300 project despite the cancellation of the same by AirAsia[2] which was 4 months later ratified by the showcasing of the FTVs, one of which was for the 160 seat variant.

The Wall Street Journal, much later in May 2015, also reported that Bombardier had been considering a longer, stretched variant of CS300, dubbed the CS500, capable of competing directly with the Boeing 737 & Airbus A320 family aircrafts in the 160-180 seat narrow body aircraft segment with the company still not having committed to the development[3].

The next key milestone for the C-Series program came in early March 2013 with Bombardier providing a comprehensive update on the program's status in form of displaying the first fully assembled Flight Test Vehicle (FTV) along with 3 other FTVs which were to be a part of the flight test phase of the program. One of the FTVs confirmed the existence of the 160 seat "Extra Capacity" variant featuring two sets of over the wing emergency exits[4]. The electrical system testing of the first FTV was also carried out along with static airframe testing & the systems tests at this stage in March 2013.

By this time (March 2013), PW1500G; the variant of the Pratt &

Whitney's PW1000G family which was to power the C-Series; had been certified by Transport Canada with the certification announced on February 2013 following an extensive 4,000 hour flight testing program for the PW1500G which originally had started in September 2010[5].

However, the PW1500G was to become another minor stumbling block for further extension of the C-Series' program schedule during the flight test phase of the program as part of its long & winding journey towards production & entry into service.

The powering up and static airframe testing in March 2013 were followed by ground vibration tests & final software upgrades conducted on the FTV1 in late June 2013, in preparation for the test aircraft's maiden flight, scheduled for late July 2013 at the time. Along with this, Safety of Flight (SOF) statements from all suppliers had been obtained and Bombardier applied for Transport Canada for the flight test permit for the FTV1[6]. Mike Arcamone, President, Bombardier Commercial Aircraft gloated at the occasion "Only five years after launching the C-Series airliner, we're approaching our maiden flight – a historic moment for Bombardier and a game-changing moment for the industry."[6]

Transport Canada granted the flight test permit to Bombardier on August 30, 2013, thereby, enabling Bombardier to conduct high speed taxi testing & flight testing on the program.

This set the stage for the maiden flight of the CS100 which was to take place on September 16, 2013 from the Mirabel Airport amid much fanfare to the highly anticipated program in the company's history.

The maiden flight ceremony was attended by the Bombardier's board members, partners, suppliers & almost 3000 Bombardier employees in an emotionally charged environment almost descending towards schmaltz as echoed by Rob Dewar, Vice President & General Manager, C-Series

Program, "This is a very proud day for Bombardier and a true validation of the C-Series aircraft's design & development, and of our extensive ground test program. "Five years in the making the C-Series aircraft's first flight is the culmination of an incredible amount of hard work and dedication from our employees, partners and suppliers around the world." [z].

The Chief Flight Test Pilot, Bombardier Flight Test Centre, commanding the FTV1, Captain Charles Ellis was elated post the return of the FTV1 to the airport after the successful flight test and he commented, "The performance of the C-Series aircraft was very impressive! We couldn't have wished for a better maiden flight. FTV's state-of-the-art flight deck was responsive and comfortable, and the aircraft handled exactly as expected. Overall, we had a very productive first flight and an excellent start to the flight test program". [z]

Similar reactions regarding the aircraft's handling & performance were to be echoed by the industry pilots much later when the aircraft finally entered service.

The much awaited maiden flight of the C-Series incidentally was also the first flight for the Pratt & Whitney's Geared Turbofan PurePower engine family following a 20 year development horizon involving $10 billion of investments by the engine maker which had announced its return to the commercial aviation market with a loud roar since losing market share to GE in the 1980s.

The maiden flight was followed by the second flight of the FTV1 on October 01, 2013. At this stage the program was running 9 months behind schedule with the maiden flight, which was to have taken place by the end of 2012, actually taking off in September 2013.

C-Series Almost on the verge of Mayday!

There was some bad news in store for Bombardier which announced in January 2014 that the C-Series program's original development schedule would have to be extended with the program needing a much extended flight test phase towards certification.

The new schedule pinned the certification to be achieved in the second half of 2015 meaning the program was going to be delayed by almost 2 years translating into extra development costs and potential cash flow issues emanating from a delayed entry into service.

However, there was more trouble in store for the already troubled program. The PW1500G engine program on one of the CS100 FTVs had an unexpected, uncontained engine failure in May 2014 which virtually brought the flight test phase of the program to a grinding halt virtually.

The same prevented Bombardier from showcasing the CS100 at the then upcoming Farnborough Airshow in July 2014 which also had an impact on the Bombardier's plans to further expand program's order book at the key industry event.

At this stage, the program was now embattled, running over 2 years delayed, burning cash rapidly over development and having gained limited traction in the program's order book owing to a tough market environment and facing scorching competitive intensity from Airbus & Boeing on a hostile terrain behind enemy lines.

The situation led to first set of casualties at Bombardier with the company announcing workforce layoffs and change of program management overseeing the C-Series which also was going to ultimately lead to the ousting of Pierre Beaudoin as the CEO and arrival of Alan Bellmare a year later in 2015.

Bombardier's order book for the C-Series had been slow to tick with the program's firm orders tally standing at 177 C-Series aircrafts. The numbers had grown only slightly from 148 firm orders at the end of 2012 amid market environment & competition from Airbus & Boeing.

Bombardier's top line, too, had been almost stagnant over the years indicating some early signs of a potential financial distress going forward as the company had been burning cash at a rapid pace on 3 development programs simultaneously.

The unexpected setback meant additional delays towards program's certification which turned out to be 4 months with the engine problem identified & resolved by September 2014 leading to the C-Series getting airborne once again with the resumption of flight testing.

Within a year, by October 2015, the C-Series had clocked around 90% of the flight test hours required and Bombardier was expecting a certification for the CS100 by late 2015 followed by an entry into service in 2016.

The CS100 received certification from Transport Canada in December 2015. However, it had come at a huge $5.4 billion in development costs so far, a tally which included a write off of $3.2 billion, booked by the company to its books as part of its Q3 financials for 2015 effectively signaling that its ship was rocking against the high tides triggering a firefight focused on saving the imperiled ship.

Bombardier; used to placing huge, long term bets leveraging easy access to stockholders and taxpayers money as a demonstrated, successful strategy for decades till the C-Series fiasco; had already burnt enough cash so far and was looking towards the taxpayers funds for bailout by funding a difficult rescue mission amid inclement weather conditions.

The first SOS sent by Bombardier was in form of a partnership offer proposed to Airbus for a majority share in return for an aerial rescue bid which, however, was turned down by Toulouse.

Canada's incoming Trudeau government & the Quebec government also received the SOS and Quebec was the first one to the site for rescue with the first tranche of CA$1 billion in bailout funds announced on the very same day of October 29, 2015 (when Bombardier first gave smoke signals) to enable the C-Series to be able to take-off.

However, this was not going to be enough for Bombardier which had actually created a black hole for itself with the C-Series. The only saving grace for Bombardier at this stage in 2016 was that the CS100 was beyond the investment heavy development cycle and was in revenue service by mid-2016, post EASA & FAA certifications and was soon going to be joined by the larger sibling, CS300; which received a common type rating with the CS100 in late 2016.

In the meanwhile, the formation of a limited partnership between Bombardier and Quebec province in July 2016; with Quebec holding a 49.5% stake in the C-Series partnership ceding operational control with Bombardier; provided another mid-air refueling to the flagging C-Series in the form of a US$1 billion cash injection. In return, Bombardier was to keep the C-Series production line in Quebec for the next two decades.

Following the successful entry into service for both C-Series birds, Bombardier drew up an ambitious production ramp-up plan for them for 2017, as per which the production was to be ramped up from 7 aircrafts for 2016 to a rate of 30-35 aircrafts for 2017.

However, sustained supplier side issues, led by continued technical troubles on the Pratt & Whitney's PW1000G engine quashed & eclipsed any remaining hopes for Bombardier to be able to push the throttle on C-Series production to expand revenues & incoming cash flow generation streams following an extremely protracted & expensive development cycle which had put Bombardier's very survival at stake by now.

These supply side issues over the PW1000G engine family, however, also had another victim in the same year, in form of the Airbus camp, and had a debilitating effect on the A320neo deliveries tally for 2017 with a large number of A320neo gliders finding themselves parked across Airbus production facilities with each eagerly awaiting a pair of geared turbofans to be able to get airborne quickly.

Bombardier, thus, was in an all out battle on all fronts amid dwindling financial resources to sustain its aggressive campaigns & escapades. Production ramp up had been impeded by the engine supply issues made worse by the decision to go with P&W GTF as the sole power plant on the program coupled with intense competition from the way larger in size, big boys of the aviation town, Airbus & Boeing, who had slashed prices on their A320neo & the 737 MAX significantly thus effectively barricading Bombardier from keeping its order book ticking over with new aircraft orders in 2017.

This pressure on order book in 2017 was coming after a robust 2016 which had been preceded by a complete washout in 2015 with no new orders at all and 61 orders coming in 2014[8]. Bombardier had booked 117 orders for 2016, its best year, at heavy discounts moving the C-Series order book in a spurt to 380 firm orders[8], including, a key sales deal with Delta for 75 C-Series aircrafts signed in April 2016 as the U.S. launch customer.

The Delta deal provided Bombardier with a much needed break through into the U.S. market apart from a boost to its share price while allaying stockholders fears temporarily with the Delta's on boarding of the C-Series. The deal, however, also invoked Boeing's ire with Delta being a majorly 737 operator with a large share of 737s traditionally maintained in its fleet.

Boeing itself had given Bombardier a major pass on a key corner of the track on the U.S. circuit just a month earlier in March 2016 by booking a deal with United for 25 of the, to be phased out 737-700s, at a rock bottom price of $22 million per airplane[9] going for the bulls eye with an almost sniper shot which just couldn't have been tenable for Bombardier on the pricing aspect. This had followed another deal between Boeing & United for 40 737NGs earlier in January 2016 (United later must be thanking its stars for having bought the 737-700 from the NG series rather than MAX).

Boeing was simply trying to keep Bombardier off the U.S. market as well as United and it was ready to do so at any cost with a 'come what may' posture, just like it did on the KC-X with Airbus, using its financial heft against a rickety Bombardier. Boeing's fear was that a determined & revisionist Bombardier could usurp 737-700's market and also shorten its remaining shelf life (expected to still have around 3-4 good years as per Boeing) while posing a strong & direct challenge to the upcoming 737 MAX-7 as well.

Boeing's this extreme undercutting of Bombardier with aggressive defensive maneuvers had Bombardier rattled and that's why it shot back sharply from behind in the super-cruise mode and got to Delta within a month with a smoky & sizzling deal for the C-Series which had Boeing fuming with full afterburners!

The next rumbling maneuver in this high speed tactical dogfight between a Maverick Bombardier and the Grandmaster Boeing came with Boeing reaching out for the referee, as usual, in April 2017, which in this case happened to be the U.S. trade representative and complained about the huge launch subsidies provided by the Canadian government to Bombardier for the C-Series which were enabling Bombardier to offer the aircraft at prices which were even below the production costs.

However, this is nothing new for the industry which is characterized by huge investment requirements followed by a long time horizon to break even that act as steep entry barriers to be surpassed for any potential, new entrant. Governments all over the world have been supporting their aerospace industries as a risk sharing partner at inception since the very genesis of flying looking to build a competent & capable industry, technology development and for employment & exports reasons. This has been a very common practice globally but still there has been a lot of cacophony over this along with its extensive & repeated usage as a potential 'smoking gun' to take down enemy aircrafts for decades.

Boeing's filing of dumping complaint was over the Bombardier's deal with Delta, for 75 CS100 aircrafts along with 50 options in the boot, which Boeing claimed that "would put its very future and that of U.S. aerospace industry at risk[10]." This was Boeing's testimony made before the U.S. International Trade Commission in May 2017 prior to its filing of a complaint with the U.S. Department of Commerce & the ITC.

Boeing also accused Bombardier of predatory pricing claiming that Bombardier sold the CS100 aircrafts at a throwaway price of $19.6 million[10] trying to undermine the market for the 737-700 & the 737 MAX-7, qualifying legally as 'dumping'.

Delta in its testimony, however, denied the predatory pricing claim raised by Boeing and clarified the selection of CS100 over 737-700 or MAX7 citing that Boeing did not have a comparable product offering and further added that these (-700 & MAX7) were 1960s era designs which didn't match the C-Series in operating economics and were thus rejected.

Boeing, however, argued for the 'injection of funding by the Quebec government towards bail out' to be treated as subsidies enabling dumping that needed to be guarded against with the imposition of 79% tariffs by the

ITC[10].

The U.S. International Trade Commission found in June 2017 that the U.S. Aerospace industry could be threatened by Bombardier's antics followed by the Department of Commerce laying the preliminary groundwork & foundations for a potentially insurmountable 'Great Tariff Wall of the U.S.A.' for Bombardier.

This was like a high intensity blast from an IED planted right under Bombardier's very own hull. Fortunately, it was a 'Double-V Hull'[5], courtesy GDLS Canada, and Bombardier survived. However, the high cost exchange ratio involved for Bombardier in effectively defending the onslaught meant another existential crisis & firefight for survival from an impending implosion and it was again on tenterhooks without an escape route from the situation.

Bombardier hadn't been able to book even a single new order for over a year between signing the Delta deal in April 2016 and Boeing's filing of complaint in April 2017 amid a strong, simultaneous continued, counter charge on the flanks from Airbus & Boeing that had Bombardier almost spinning out on the track.

The situation was getting even more complicated with the entire U.S. trade defense batteries now targeted & locked on the C-Series meaning changing coordinates also would not have helped and it was time for retreat.

Bombardier literally was in dire straits now facing a multitude of High-G forces acting simultaneously having created a state of G-LOC & hypoxia with production ramp-up not doable over supply issues, a stagnant order book with cut throat competition challenging profitability of C-Series, dwindling finances with mounting debt levels, credit rating downgrades,

---

[5] Double-V Hull is a trademark of General Dynamics Land Systems (GDLS)

tumbling stock prices, shareholders ire & fury grilling the top management and on top of that a rapidly closing door to the indispensable U.S. market. It was literally a ticking time bomb waiting for the implosion…

While Bombardier was embroiled in the middle of this raging inferno, there appeared a rescue chopper on the distant horizon in August 2017 presenting a much needed & vaunted glimmer of hope to those on the imperiled deck. And it was an Airbus make, the heavy duty H225 Super Puma, the champion of the fiery trails (except its crash track record), in Search & Rescue (SAR) configuration and the crew spoke in the same language, French. The French indeed had arrived to the rescue, once again!

However, this time round, it was for economic reasons as well rather than merely geopolitical. Lowering the hydraulic rescue hoist and marking a tectonic shift in the industry landscape, Airbus announced in October 2017 that it had taken a 50.01% stake in the C-Series Aircraft Limited Partnership (CSALP), owned by Bombardier & the Quebec government so far, which were to now own 31% and 19% of the entity, post Airbus on boarding.

Airbus announced that it would bring its massive relief forces on board and would put its procurement, sales & marketing and customer support muscles behind the beleaguered C-Series in order to stabilize the rocking ship and steer it back on course.

There were indeed several pluses for Bombardier from the deal. Airbus on-boarding would have boosted customer confidence in the C-Series and would have come as a shot in arm for the order book. Additionally, Airbus supply chain heft would have sorted the production costs by streamlining things and putting the C-Series on the long path towards profitability. The Final assembly line for the C-Series was to be retained within Quebec while it was also going to be produced at the Airbus' latest production facility based in Mobile, Alabama to circumvent the tariff barriers.

For Bombardier, it was akin to a tortuous journey to hell followed by a dramatic comeback ride on the way back. However, the landing, back on to the ground for Bombardier going forward, was still going to be a rough one given mounting debt levels and no upfront cash payout from Airbus for the deal. Boeing clearly had missed out on the irresistible deal having hung up on the distress call which had been received at Seattle as well!

For Airbus, it was nothing short of a steal & a masterstroke with the company getting effective control of the 21st century's first operational, clean sheet aircraft incorporating state of the art technologies without any upfront costs or investments.

Airbus also had just established its fifth global narrow body production facility at Quebec, Mirabel and its second one in North America. The C-Series, which was now rebadged as the Airbus A220, also had a good strategic fit with Airbus' narrow body line up at the lower end of the portfolio.

Further, it gave Airbus another very capable & unmatched stallion to lead the charge at Boeing's bastion, for which Boeing had no answer as it lacked a similarly capable mustang in its stable. Also, the decision for the Mobile ranch meant tariffs were meaningless and the ongoing duel was going to be fought out & settled in the market place instead. Airbus CEO, Tom Enders exclaimed: "This is a win-win for everybody! The C Series, with its state-of-the-art design and great economics, is a great fit with our existing single-aisle aircraft family and rapidly extends our product offering into a fast growing market sector" [11]. The road sign at the fork read: 'Advantage Airbus!'

John Leahy summed up the entire dramatic sequence from the playbook perspective succinctly in an interview much later after his retirement. He said: "I always said it was a cute little airplane, and it is (referring to C-Series). We were competing against it for years, rather successfully. I think so did Boeing, so I am shocked beyond belief that Boeing was trying to kick a dog when he was down. If Boeing had totally ignored them, they probably would have rolled over and died. Why they had to go pressing the Trump Administration for 300% tariffs is beyond the pale. I think a lot of reports that they (referring to Boeing) forced them into our hands are probably true[12]".

Boeing clearly had been saber rattling while having gone in for the overkill and had overreacted by almost treating a $200 worth UAV's incursion & intrusion into its airspace as an invasion and had tried to take it down almost with a $3 million worth Patriot surface to air missile system!

Boeing, thus, eventually forced Bombardier to go across the Atlantic towards its archenemy's zone which simply lapped it up & made it a part of its already formidable arsenal adding further depth & range to the existing firepower.

The Mustang, thus, had landed & settled down in the French Alps and Boeing was now busy again, this time preparing for a counter attack strategy to what had just transpired.

By December 2017, the Department of Commerce, nonetheless, had slapped 292% duty tariffs in its final ruling palisading the U.S. airspace for Bombardier. However, all of it did not matter anymore and based on Canada's filing of a protest in the WTO against tariffs in January 2018, the USITC ruled that the U.S. industry was not threatened and overturned its earlier ruling & the imposition of duties!

Boeing's entire think tank had been engrossed in devising its own plan to counter a growing Airbus in North America and the war room came up with the plan to go southwards towards South America and lap up Brazil's Embraer; Bombardier's stark enemy in the regional aircraft domain;

echoing and much in accordance with the longstanding, early 19th century era Monroe doctrine!

Embraer, however, was no Bombardier and was not born with the silver spoon either which could have granted it the special deal hunting or betting privileges. Embraer, used to operating with a relatively much safer & conventional playbook traditionally, thus, did not have a clean sheet commercial aircraft program to boast of. However, what it had was a large pool of proven engineering expertise available at a highly competitive operating cost base from a relative perspective, ready & all geared up to be deployed for any potential mission profile at very short notice.

Boeing; which had for long been contemplating a 757 replacement, had been leaning actively towards the New Mid-Market Airplane (NMA) in the 21st century and knowing well (hopefully) that it eventually will have to decommission the 737 someday in the future by creating a worthy successor in time; seemingly had this 'game plan' in mind for the yet to be on boarded Brazilian troops.

Boeing announced its own potboiler master plan in July 2018 of joining forces with Brazil and form a JV based strategic partnership with Embraer with the latter's commercial aircraft business at the core.

As per the plan, Boeing was to acquire an 80% stake in the Embraer's commercial business worth $3.8 billion[13], as per the MoU signed, and was to manage Embraer's commercial aircraft portfolio spearheaded by the E1 & E2 regional Jets.

The plan had sound undergirding with a potential for almost $150 million[13] worth of cost synergies annually by third year & a good overall strategic portfolio fit. However, unlike Airbus, Embraer was devoid of a C-Series kind of trump card or firepower given the E2's were a mere re-engined & re-winged derivatives of the E1.

Embraer's E1 Jets were ageing and the E2 jets, had not gotten off to a good start post take off, with the exception of the E195-E2, and Embraer had erred big time on the E175-E2 as it is heavier and not U.S. scope clause compliant as Embraer was expecting long anticipated revisions in scope clauses to have materialized by its EIS, thus, effectively keeping it out of the U.S. skies.

Embraer was hoping that flying under Boeing's wing would give the E2 birds the much needed sail into the wind to overcome the headwinds they faced with an indolent order book. Additionally, Boeing & Embraer were to jointly; under a separate JV; develop newer markets globally for the Embraer's defense portfolio, led by the KC-390 military transport aircraft, a tactical airlifter which had not been able to penetrate many markets beyond Latin America & parts of Europe, linked directly with Brazil via their colonial legacy.

The Airbus vs. Boeing rivalry & bitter acrimony thus was going to have a new war theatre, the lower end of the commercial aircraft market and the regional jets territory along with new state of the art weaponry to wage battle campaigns aligned with the detailed war doctrines of the old masters.

This almost was going to lead to further expansion of the territory & terrain controlled by the duopoly, like two entities on either side of the iron curtain, in the commercial aviation landscape across segments ranging from entry level to the very top end of the spectrum marked by the largest wide bodies. The

Boeing-Embraer deal, however, was still subject to shareholders & regulatory approvals and was expected to close by the end of 2019. Embraer, under its CEO Michael Slattery, kept working diligently over the following years since deal announcement towards carving out its commercial business from the remaining portfolio while the active pursuit of regulators & aviation authorities globally for approval was done. By the onset of the COVID-19 pandemic & the grounding of commercial aviation in March 2020, the deal had secured all approvals, except from the European Commission, which had sought additional modalities.

However, in a surprising development in late April 2020, Boeing almost shocked Brazil and the entire aviation world by announcing that it was retreating from the deal at the eleventh hour in an abrupt & unexpected volte-face over some contractual conditions.

This was after having courted the Brazilians for almost 2 years drawing Embraer's verbal chain gun of fire & fury in retaliation, much along the expected lines. The reason cited by Boeing ostensibly in its decision has been, "Embraer did not satisfy the necessary conditions[14]" with Boeing seemingly leveraging the contractual conditions as a cover to make its exit from the agreement in an 'altered' post COVID-19 world.

Embraer, having already incurred $200 million in charges towards the carve-out of its commercial aircraft portfolio in the books which shrunk the bottom line, was justified in its anger at this unexpected jilting & setback.

Embraer, firing its 120mm mortar rounds in strong retaliation, targeted Boeing heavily with scathing comments in its company statement with the first round exploding as: "Embraer believes strongly that Boeing has wrongly terminated the mutual transaction agreement and that it has manufactured false claims as a pretext to seek to avoid its commitments to close the transaction & pay Embraer the US$4.2 billion purchase price[15]."

The second round landed in the backyard, "We believe Boeing has engaged in a systematic pattern of delay & repeated violations of the MTA (master transaction agreement) because of its unwillingness to complete the transaction in light of its own financial condition and 737 MAX and other business and reputational problems[15]". The final round was bang on target:

"Embraer will pursue all remedies against Boeing for the damages incurred by Embraer as a result of Boeing's wrongful termination and violation of the MTA[15]."

Boeing, in an evidently poor tactical display of a sequence of aerial aerobatic maneuvers, in front of a fully packed house of spectators, at first, overreacting to Bombardier's antics with its impulsive retort which ultimately added to its archenemy's overall firepower and made it even stronger in an area where it already had an upper hand and where Boeing's evident weakness also lay.

This was followed by an equally impetuous, half baked Embraer potboiler, which might cause the company another set of financial cramps going forward apart from adding to a long series of self-created mishaps on the road, very frequently travelled and simply providing newer ways to tart up scoops making for interesting press coverage for an ever so hungry media.

This series of impulsive, ill planned, unpracticed, uncoordinated & poorly executed maneuvers only have created more structural stress for an airframe seemingly already past its prime indicative of the urgent need for a critical reassessment of the structural reliability, performance & continued airworthiness for which a litmus test was soon going to be headed its way…

Bombardier's hard landing post the handover of the command of the C-Series to Airbus had been much along the expected lines given the mounting debt problems which ultimately led to warranting of the structural dismantling of the Bombardier Empire, built so meticulously over decades.

Bombardier sold off assets in piecemeal way & exited its entire commercial aviation portfolio after C-Series by the end of 2019 with the CRJ portfolio bought by Mitsubishi (June 2019), flight training business going to CAE (November 2018), Aero structures business sold to Spirit Aerosystems (October 2019), De Havilland & Q400NextGen turboprop line going to Longview Aviation (November 2018).

2020 brought more divestments led by the jettisoning of the Bombardier's remaining stake in the A220 program to Airbus. The last bastion to fall in the Bombardier's industrial empire came with the disposal of the transportation business worth $8.2 billion in a mega deal to former rival Alstom in February 2020 leaving Bombardier with only the business aviation portfolio as a solace in the name of being an aerospace player. All this had to be done by the CEO Alain Bellemare, as the debt levels had gone up to $9.5 billion for Bombardier by March 2020 amid a stock price in a virtual freefall mode. Bombardier, thus, at the end of it all is just a miniaturized shadow of its former glorious self!

However, knowing now that the COVID-19 pandemic was going to hit the industry & the world like a tsunami, Bombardier's ability to withstand this kind of an impact & still stay afloat with the huge looming debt levels it had accumulated without having reincarnated itself into a much smaller self too is again highly questionable.

This was a classic example & a methodical action plan possibly titled 'how to dismantle an industrial empire' on which an entire tome could be penned down for a bestseller or a detailed script followed by a vivid screenplay could be written for an adventurous epic for celluloid.

Bombardier had been operating & winning successfully over decades with its proven strategy & business model of 'buying cheap, turn it around and reap huge profits' betting on stockholders money with government assistance for long term growth. This had propelled Bombardier's rise from a snowmobile manufacturer, till early 1970s, to an industrial giant by the turn of the century.

However, with C-Series, Bombardier had erred by straying & deviating significantly from its typical & longstanding model. Here, the model involved investing huge sums to sneak its way into a well contested market with huge entry barriers and well entrenched, much larger players with huge financial & marketing resources.

Bombardier had gotten itself trapped effectively by looking out & aiming for the no man's land between the two arch-rivals, literally para-dropping its way into the mine strewed piece of land by taking the leap of faith and ultimately becoming a well targeted casualty of the routine crossfire and exchange of 155mm artillery shells between the two traditional arch-rivals.

Bombardier, in fact, earlier in the late 1980s, too had a narrow miss on this very same trajectory when it had considered venturing into the automobile business riding on Japanese Daihatsu which would have effectively positioned Bombardier right in the middle of the Ford & GM's cross border shelling zone and the results would have been equally catastrophic!

This had to happen and was waiting to happen given Bombardier's betting approach to growth driven by the management's strategic overstretch coupled with an unflappable exuberance while placing huge bets with limited accountability for potential repercussions & fallout. That adrenaline rush sometimes becomes an addiction with the aftereffects further compounded by that very unflappable exuberance which further loosens one's grip on the hardcore reality leading to a hard landing ultimately accompanied by casualties…

Little did Bombardier realize that with the C-Series it was actually boarding a kamikaze, which after inadvertently hitting Boeing as the ultimate target by default, would go on to put its own barn on fire in the auto-pilot, self destruct mode!

"He who relies solely on war like measures shall be exterminated; he who relies solely on peaceful measures shall perish!" –

Sun Tzu – The Art of War

Chapter – 10

THE 'LONG-LEGGED' STALLIONS, SHIFT IN 'BALANCE OF POWER' AND A COUP IN THE 'MIDDLE KINGDOM'

"Opportunities multiply as they are seized". – Sun Tzu, The Art of War

The 'Battle of C-Series' marked a major victory & a key milestone for Airbus in its traditional and over 5 decades long, all out war with Boeing for market supremacy. The A220 comes with a very strong positioning & proposition with a significant strategic & competitive overmatch at the base of the spectrum for Airbus' narrow body portfolio giving great company in the formation flight to the A320neo, which is already well positioned in the orbit as the nucleus & core of the portfolio.

The A320neo aircraft family, starting early with a first mover advantage, has maintained & even extended the lead by a large margin having piled up laps on the 737 MAX and clearly has given it a harrowing time across circuits. The victory has come like a nitro boost for Airbus injecting a significant dose of oxidizer propellant to up its game further while gearing up for the next big battle with Boeing.

With the lower end and core of the narrow body spectrum settled, the next battle for Airbus to be won was for the Middle of the Market, the 150-220 seats segment; a territory long dominated by the Boeing 757, since its entry into service way back in 1983 and final disembarkation from the assembly lines in 2004.

The geopolitics of the middle of the market has been along the classical lines marked by the Boeing's 757-200's clear hegemony sustained for long followed by great powers struggles with the succeeding contenders (post the change of command at Boeing) slugging it out in the battlefield for supremacy with the advent of 737-900 & 737-900ER vs. A321-200 and later the neo vs. max contest.

The debacle of the A380 and the success of the 787 were accompanied by the Boeing induced market shift towards point to point model with the opening of multiple new city pair routes. However, some city pairs were long range as well as thin, in terms of air traffic, and thus were not easy to fill and did not require the passenger capacity or the long legs offered by the twin aisle 787.

The smallest 787 variant, the 787-8, typically carries 242 passengers flying over a 7,355nmi range. The single aisle 757-200 (smaller of the two 757 siblings), positioned right in the middle of the market, carries 200 passengers over a range of around 4,000nmi[1], which has typically remained more or less as the goldilocks equation and the centre of gravity for the middle of the market. The larger 757-300 typically carries around 243 passengers over 3,400 nautical miles. The 757 maintains the same cabin cross section as the older 707 with a common type rating certification with the 767 wide-body, developed concurrently.

The 757 program; starting out as a replacement for the 727 trijet with a

80% increased fuel efficiency[1] and following its initial tryst with turbulence early into service characterized by a 3 year long orders famine; made a strong charge later in its career post changes in regulations and the onset of the waves of globalization. The 757 proved to be quite popular with the airlines which have leveraged the versatility offered by the aircraft to the max by putting it into service extensively over longer, mid-range transatlantic routes since the modifications to the extended range twin engine operations (ETOPS) regulations came in the late 1980s.

Delta and United have remained as the largest operators of the 757 in the U.S. market. The 757-200 especially has proved to be most popular with Boeing having rolled out around 900 of them by the time production ended in 2004[1] as a fallout of the post 9/11 U.S. aviation meltdown with the baton ultimately relayed to the old warrior, 737-900/-900ER; the largest variant of the 737NG family.

The Airbus heavyweight, which has traditionally somewhat come close to be able to have the gumption to lock horns with Boeing's 757 in the segment, has been the A321, the A321-200 to be precise.

Seeing launch in 1988 & deployment in 1994 as the first stretched derivative of the A320 family and the first A320 family aircraft to have German antecedent & passport (courtesy the Hamburg final assembly line), the first variant A321-100 (without AFTs) was a no match for the 757-200 in terms of range and soon gave way to the advent of the heavier & longer range A321-200 variant for duel with the launch in 1995.

The A321-200 has a passenger capacity of 170 to 220 passengers with an operating range of 3,200nmi[2], which still was somewhat short of the 757-200's around 3,915nmi nevertheless but was a fitting match for the later mere mortal contenders fielded by Boeing, the 737-900/-900ER.

Airbus has been actively promoting the A321's cabin width, at 3.70

meter[2] (as against 757-200's 3.54 meter & 3.53 meter for the 737), which is the largest compared to any direct competitor, as per Airbus, translating into the grammar of passenger comfort with the 18 inch wide seats and when combined with the common type rating with the other members of the A320 family, also into the logic of operating economics for the airlines.

The same has been reflected by the A321's successful score card with over 2,000 strong aircraft deliveries so far and a slightly over 3,000 strong firm order book position (including neo variants), as of the end of year 2019, as per Airbus with American leading the airlines pack in maintaining one of the largest fleet of A321s.

The numbers gain further significance when compared to the 757's relatively much longer (over two decades long) production run ended in 2004 that rolled out 1,050 aircrafts (including 80 freighters) with over 90% of them being the 757-200s.

The key point here is that since its entry into service in 1983 till the A321-200 was launched in 1995 and took to the skies in 1997, a long span of almost 12-14 years, the 757-200 was unchallenged and had a virtual monopoly of the middle of the market as the sole heavyweight leveraging its inherent advantages of the unique combination of being a long range-high (200+ seat) capacity narrow body capable of withstanding rigors of transatlantic deployment along with a decent take-off performance in hot & high conditions bundled with the fact that it could fit into the same gates at airports as a typical 737.

In fact, Boeing showcased the take off capabilities of the 757 with a stunt in 1991; marked by a single engine take off & landing from the Gonggar Airport based in Tibet; situated at an altitude of 11,600 foot, from a box canyon shaped airfield set against the backdrop of 16,000 feet+ high mountain peaks[1].

The airlines could either press it in service for shorter, thin transatlantic routes or fly it on more dense domestic routes with the 757-200 providing them a lot of flexibility and versatility in terms of operational scope & mission profiles.

The 757, following the ETOPS regulatory changes in 1988, still had around 7-9 strong years all alone to itself as the sole flag bearer in this segment given that none of the wild ponies from the DC stable were any match for it in terms of this golden equation of operating range, payload & operating economics.

Another contender which was anywhere close to the 757 or somehow in the vicinity in the 1990s, coming from the East, was the Tupolev's Tu-204 commercial aircraft family which was clearly modeled on the 757 with similar specs and a 200 passenger capacity albeit with a much shorter range of 2350 nautical miles.

The Tu-204 also featured fly by wire controls and did not need the wing ant-icing system (ratified by the EASA certification). The Russian grand aviation plot of the 90s, however, remained more or less a pipedream and stayed confined to the domestic market with only 86 aircrafts ultimately produced.

The post 9/11 carnage in the U.S. aviation market drove the airlines away from 757 towards smaller narrow body birds, the 737 & the A320, with Boeing's freighter market move (it's first one) in search of a potential secondary market for the 757-200, with the 757-200SF freighter variant, too, turned out to be unpromising.

Boeing also contemplated developing a further longer range variant of the 757 at the turn of the century, the 757-200X with very long legs, with a range crossing the 5,000nmi threshold[3]. The additional range was to come from the auxiliary fuel tanks held in the aircraft's belly while structural reinforcements were to be borrowed from the 757-300.

The strategic rationale behind the 757-200X was to create new city pairs by the catalyzing the further fragmentation of the North Atlantic market on thin routes leveraging the extended 5,000nmi range to the hilt which would have enabled the -200X to cover almost the entire Western Europe from the JFK on the U.S. East Coast while also enabling potential incursions far deep into Eastern Europe almost to St. Petersburg on a 12 hour grueling body shaker.

However, the -200X was shelved with no orders forthcoming at the time but nonetheless the idea of a narrow body capable of flying as long as 5,000nmi with AFTs was revolutionary & clearly ahead of its time and its time was going to come a decade and a half later but not for Boeing to be able to ride it…

The 757 program; with the market hinting towards shift to smaller narrow bodies, 757-300 failing to take-off, 757-200SF headed nowhere and the 757-200 finding no direction to head towards for future; was wound up in late 2003.

In the post 757 world, the airlines now either had to shift downwards towards smaller, mainstream narrow bodies or had to shift upwards towards the 787. However, the discontinuation of the 757 production still left a deep void & a kind of vacuum in the marketplace which somehow stayed with the status quo situation for a long time since the relegation of the command to the 737-900/-900ER, which (-900ER) entered service in April 2007.

This was given that the -900/-900ER were still way below & unable to match the 757's above-par maverick of the narrow body pack kind a capabilities threshold with their lower operating ranges & capacities with the -900ER, even with the AFTs in its belly could only cover 2,950 nautical miles at max with a MTOW of 187,700 lb as against 757-200's 3915nmi[6] with a MTOW of 255,000 lb.

The A321, with its 3,200nmi range along with a MTOW of 206,000 lb, too was a middleweight at best against the 757-200's heavyweight class but still had a slight edge over the 737-900ER. The next echelon, the 787, was in a class way above with its 7,000nmi+ starting range with story being the same for the Airbus camp with the A330-800's even longer legs capable of covering 8,000nmi+.

This was when the 757-200s were powered by a pair of the bygone era; fuel thirsty RB211s and the PW2000s as engine options from Rolls Royce & Pratt & Whitney respectively. The newbie narrow bodies on the contrary were powered by the CFM56s & the V2500s as options.

The Maverick 757 narrow body, thus, was never replaced in essence since its departure and the market's 'Top Guns', especially; the U.S. based major carriers have continued to maintain their 757 fleets in active service. As of 2018, around 666 Boeing 757s (out of the 1050 produced & delivered over 23 years[1]) were still in active service across variants; led by Delta, FedEx and United; with the trio collectively accounting for almost 50% of the in-service fleet of 757s[4].

There was, however, another variable within Boeing, though not directly related, which moved much in tandem with the 757's life cycle, especially, the sharp change in its fortunes at the turn of the decade in the 1990s; and that variable was the career graph of Boeing's eighth Chairman & CEO Phil Condit.

Condit had spent over 8 years of the initial part of his long career within Boeing on the 757 program under different capacities and had a major design influence on the 757 program. His rise to the very peak of the

---

leadership ladder within Boeing in the mid-1990s was accompanied by the 757's summiting of its career peak at the onset of the same decade and his sudden exit from the scene in late 2003 was again in an uncanny way reflected by the 757 with the program's last set of orders for 7 aircrafts coming in 2003 followed by the production shut down decision manifesting within a year of his departure in 2004!

The 'Middle Ground Theory' and a 'New Bird' joins the Flock

Airbus, following the launch of the A320neo in December 2010 & its super successful take off, wanted to go a step further for another competitive edge & first mover advantage over Boeing with the launch of the re-engined version of the larger A321 as well, i.e. the A321neo.

The A321neo had the same pair of new engines, as the A320neo, with the option of choosing amongst the Leap-1A from CFMI and Pratt & Whitney's PW1100G kept intact.

The other key changes made by Airbus on the A321neo included the slight increase in Maximum Take-Off Weight (MTOW) from 93.5t on the A320neo & the A321ceo to 97t on the A321neo with structural reinforcements to the wing & landing gear (driving a slight increase in empty aircraft weight of around 1.6t on the neo) which along with configuration changes with the creation of second over wing exit & the introduction of Airbus Cabin-Flex (ACF) layout enabled the A321neo to include an additional 20 seats.

This led to increasing of its maximum seating capacity in a single class layout to 240, up from 220 on the original A321, translating into additional revenue potential for airlines along with a double digit percentage boost to the overall fuel efficiency per seat with newer engines and incorporation of sharklets.

The A321neo, incorporating the German 'Feinmechanik' from the Hamburg line, had its maiden flight in February 2016 (while the first 737 MAX had just taken off for its first test flight in Jan 2016) and received its clearance to get airborne with the P&W's GTFs with EASA type certification coming within the same year in December 2016 followed by the CFM LEAP powered variant in March 2017 paving the way for EIS. The first LEAP powered A321neo was delivered in May 2017 to Virgin America.

The A321neo's order book, in the meanwhile, had been moving steadily (despite a slight order intake blip in 2016 led by the wide bodies) with airlines profitability consistently having stayed in the black zone for 8 years & staying in the rally mode after peaking in 2015 led by the U.S. carriers and their business confidence index headed skywards.

The A321neo thus had gotten off to a great start except for the teething troubles being faced by the PW1100Gs, which had sent original aircraft delivery schedules for 2017 across both Airbus & a Bombardier (operating in firefighting mode over aircraft delivery targets) literally for a toss causing severe disruption & chaos.

However, the operating range for the A321neo was still 400-500nmi short of the 757's just under 4,000 nautical miles in the middle of the market segment, which led to the creation of the next iterative derivative of the A321neo under a new game plan by Airbus to outflank Boeing & wrest control of the middle ground as well.

The Missing Piece of the 'Middle Ground' Puzzle is Found – In form of the A321LR

The A321neo and its subsequent market success drove Airbus towards pushing the game even further by going deeper into the Middle of the market, a traditional Boeing stronghold, to unlock a sizeable growth opportunity for its self by disrupting & altering market dynamics fundamentally.

Despite retirements, the U.S. market still had a sizeable fleet of older 757 aircrafts in service, around 666 as of 2018[4], as the airlines needed replacements but there were none as Boeing's efforts towards replacement activity, anchored on a new clean sheet design, mostly remained confined to market studies, papers & plans and did not materialize into any concrete action (more on it a little later) as Boeing could not freeze market requirements into any concrete shape or form given $10-$15 billions at stake.

An untapped 700-800 aircraft replacement market, with 100 of them operating dedicatedly serving the transatlantic routes; that could be harnessed based on a low risk product strategy centered on a proven, in-service aircraft program; was nothing short of a treasure trove waiting to be dug out and almost worthy of creating a virtual Aviation Gold Rush of the 21st century in an almost wild west setting.

For the Airbus camp, it was time for vengeance & to turn the tables on the original master of the middle ground by taking a cue from Boeing's original, self authored playbook with a low-cost & low-risk, derivative pathway pivoted on the proven & in-service A321 aircraft platform along with incorporation of ACTs/AFTs for the additional range at minimal risk & little additional strain on airlines budgets with the common type rating for the A320 family taking care of the rest.

Boeing had created an extended range variant of the largest member of the 737NG family, the 737-900, in the late 1990s, dubbed the 737-900ER, sporting three AFTs (Auxiliary Fuel Tanks) to become a stopgap successor to the 757. Boeing has used the AFTs tactically & extensively across aircraft programs for long, looking to create that additional maneuvering margin & kinematic performance leveraging thrust vectoring for the high octane dogfights (like the 767-300ER vs. A300).

The concept of AFTs/ACTs provides tremendous operational flexibility to airline operators in terms of expanding the range of their aircrafts across varying payload conditions or increase payload on fuel limiting ranges without impacting bottom lines significantly and without taking the aircraft off service for long as the tanks can be installed & activated with just an 8 hour turnaround time, as per Airbus.

With this game plan at the core, the 'Battle of the Middle Ground' was set up & ready to be waged and the bugle sounded the Airbus charge. The A321LR (initially dubbed A321neoLR), the longest range variant of the A320 family with a 97t gross weight & capable of covering 4,000nmi (100nmi more than the 757-200W used on transatlantic routes & 400-500nmi more than the A321neo) with 3 ACTs & 164 passengers onboard was launched in January 2015 with Air Lease Corporation as the launch customer, following a successful marketing campaign that had been initiated a few months back in October 2014.

The A321LR had an unbeatable value proposition in form of 27% lower trip costs & 24% lower per seat costs than the ageing 757 and its overall operating costs were going to be 25% lower than the 757, as per Airbus[5].

The underlying strategic intent behind the launch, to establish a clear portfolio overmatch in the segment, was also outlined by the then Airbus EVP Sales & Marketing, Kiran Rao, in an interview. He had said: "The

A321neoLR will be a very unique aircraft when compared with the [Boeing] 737-9, which has no development legs left in it[5]."

The A321LR plan had sound conceptual underpinnings for long & thin transcontinental routes with a mature, proven & well established narrow body aircraft variant with extended legs covering the extra miles at competitive operating costs without much financial risk. Consequently, the A321LR would have been a great fit for the U.S. East Coast to Western Europe, North America to South America, Europe to Africa and South East Asia to Australia kind of routes. The LR was scheduled for entry into service in 2018 following into the footsteps of the A321neo, which was to take off 2 years earlier in 2016.

The Boeing stable, on the other side of the Atlantic, did not have the advantage or privilege of a similar, proven workhorse with such extended legs and on the track actions matter & create the decisive difference and not the pedigrees. Even the longest range variants of the 737 MAX aircraft family, the MAX 9 & MAX 10, max out at the 3550nmi range and effectively are no match for the A321LR's 4,000nmi.

The LEAP-1B engines, which power the 737 MAX family, have the smallest fan diameter and lowest thrust output, of all the LEAP engine family variants developed so far, given the 737 airframe limitations, thus effectively closing the option of mounting larger engines even as an option.

The Airbus, with the A321LR, thus, had created its first overmatch over Boeing in the Battle for the Middle of the Market at a mere fraction of a cost of what it would have potentially cost Boeing with a clean sheet program of its own, as it was the only option left for it.

Spearheaded by the A321LR, the A321 program's order book got a tailwind boost and started cruising effortlessly at high sub-sonic speeds with the commercial aviation market already in the super-cruise mode.

By the end of year 2017, the A321neo program order book (including the LR variant) had even surpassed & overtook the original A321 with an order tally standing at 1,920 aircraft units (led by Air Lease Corporation, Wizz Air, American, Delta & Air Asia with their three digit orders) accounting for almost 14% of Airbus' cumulative order book position for its A320 aircraft family.

This was much in contrast to the original A321ceo which had gasped out at 1798 orders with at a 12%+ share of the A320 order book[6]. Collectively, the introduction of the A321neo & A321LR had boosted the overall share of the A321 program (including all variants) from the 12% to 13% range to almost over a quarter (26%) of total orders for the A320 family within a span of just half a decade!

This was like a page right out of Sun Tzu's Art of War, the classic military treatise on strategy & tactics, in a perfect & flawless real world action & execution, just as the iconic general had prescribed; leverage your strengths, exploit your enemy's weaknesses and fight on a terrain that favors you but is unfavorable for the enemy; and seeing it in perfect execution must have made even the astute & legendary military general take notice from up above & chuckle..

The A321LR received joint EASA/FAA certification in October 2018 (with an up to 180 minutes of ETOPS capability over any transatlantic route) and entered service a month later in November 2018. The A321LR has enabled airlines to open multiple new city pair routes based on the operational capability & flexibility provided by the long legged aircraft. Many airlines globally have shifted to larger narrow body aircrafts over the years for these thin, transatlantic routes.

The longer range A321neos surely are going to provide wings to the growth ambitions of many LCCs by enabling & thrusting them to transcend from merely being domestic or intra-region players towards the larger, inter-regional power play going forward.

For instance, TAP Air Portugal has been in the process of acquiring around 15 A321LR aircrafts from Airbus which are going to be used on multiple transcontinental routes going forward led by the Porto, Portugal to Newark, New Jersey route.

The other transcontinental route being considered for TAP's A321LR is going to be between Lisbon in Portugal and Belem in Brazil which is going to be the first route for the A321 between Europe and South America. British Airways is currently using its A318-100 for the transatlantic flight connecting JFK, New York and London covering a distance of 3,017nmi in a span of 7 hours and 20 minutes seating 32 passengers in an all business class seating layout marking one of the rare successful deployments of the A318.

The Philippine Airlines is using its A321neo for the longest flight connecting Manila, Philippines and Sydney, Australia, a distance of 3,380nmi covered by the aircraft in 8 hours and 20 minutes. Leading Indian budget carrier, Indigo, received its first A321neo in November 2019 and the airlines plans to expand its operations to include medium haul international destinations leveraging the new aircrafts with Istanbul announced as one of the first new international destinations to be covered by Indigo from India.

The increasing trend towards shift to these longer range narrow bodies may turn out to be grueling for passengers who have traditionally been used to wide body comfort on these routes while it is likely to give a boost to the airlines in their eternal quest for an ever higher profitability.

Airbus, however, having already thrown down the gauntlet was still not settled with the score line so far and was still yearning for more, that little

'Extra' which could set the neo apart from the ordinary and be a true game changer! And it was very much in the offing.

A 'Game Changer' Neo Goes the Extra Mile: The A321XLR and a Dynamic Altered Fundamentally

The knockout punch in the middle weight class from Airbus, sealing the fate of the one sided contest, came in form of the launch of the longest range variant of the A321neo, designated aptly as the A321XLR, for the 'extra long range' it possessed and sounding the death knell for the in-service Boeing 757-200 fleet.

Airbus had been considering the development of an even farther range variant of the A321LR (as the next iterative evolution) ever since the launch of the A321LR in early 2015 looking at somewhere close to the 5,000nmi threshold and had been in talks with airlines to explore the market. The plan was to further increase the Maximum Take-Off Weight (MTOW) of the A321LR up to 101t (223,000 lb)[7] to be supported by a further structurally reinforced landing gear while keeping the same wing & engine thrust output with the optimization of the wing trailing-edge flap configuration.

The additional fuel carrying capacity for the range extension came from a permanent Rear Centre Tank (RCT) which became a force multiplier & game changer of sort with its remarkable 1-2-4 characteristics. The single RCT, by comparison, is equivalent of 1 ACT in weight, takes the space equivalent of 2 ACTs and holds as much fuel as 4 ACTs combined[8]. Additionally, Airbus could fit an additional forward ACT for additional fuel.

Image 10: Airbus A321XLR

Image Source & Credits: Airbus SE

Air Canada became the ultimate catalyst in the process of crystallization of the A321XLR, in its final form & specifications as the A321LR derivative, with a media statement in January 2019 that the group was considering the use of narrow body aircrafts for long range transatlantic routes[2].

The group further averred that it was considering contenders from both Airbus as well as Boeing. By June 2019, Le Bourget was ready for the formal launch ceremony of the A321XLR at the Paris Air Show 2019 with Middle East Airlines becoming the launch customer. The XLR variant is configured for an operating range of 4,700nmi (15% more than the A321LR and 20% more than the 757-200) with a 30% comparative lower fuel burn per seat (than the 757-200) while carrying 180-220 passengers onboard[7] with the entry into service scheduled for 2023.

The A321XLR; with its unbeatable combination of role versatility, operational flexibility, operating economics with commonality factor & the widest cabin in the narrow body segment; proved to be a blockbuster hit at the launch event at the Paris Air Show 2019 with the XLR virtually stealing the show as the event showstopper having booked 191 aircraft orders by the end of the 7 day event, which included the industry heavyweight, American Airlines, which enplaned the program at launch stage itself by placing an order for 50 aircrafts to replace some of its in-service fleet of 757-200s as the rattled & dazed team of Boeing sales executives watched from the event sidelines. Other hotshots to join the party included Qantas & ILC.

Robert Isom, the American Airlines President, exclaimed & elaborated in a vodcast (video presentation) series for American team members about the rationale, pivoted on operational flexibility, behind the decision, "It costs a little more for these aircraft but this aircraft has much greater utility for us in the long run giving us 'the ability to fly farther' and 'to be more efficient, take complexity out of the operation'. The A321XLR offers American the chance to "open up some new markets that will help the network…. For us, when we take a look at the opportunities, whether it's TransCon opportunities, or close-in Europe or close-in South America, we can configure this aircraft anywhere in between[10]".

By May 2020, the share of A321neo orders (including LR & XLR variants) in Airbus cumulative order book for the A320neo family had shot up to around 46% and accounted for 22% of the A320 family's cumulative order intake, as per Airbus, which had exposed gaping holes right in the middle of Boeing's product portfolio and had exploited them impressively in a feat of strategic legerdemain!

The entire industry had been expecting the spectacular aerial display of fireworks lighting up the sky at Le Bourget in form of two back to back program launches, led by the Airbus A321XLR launch and Boeing's long awaited announcement of the launch of NMA (New Mid-Market Airplane) program (with 220-250 seats & 4000nmi+ operating range) at the Paris Air Show, as a long overdue replacement for the 757-200, which had been relinquished with its succession plan rusting on the backburner for over a decade & a half since the 757 program shut down in late 2004.

However, much to the industry's disappointment, they only got to view the fireworks display from Airbus and what actually greeted them from Boeing camp was a complete radio silence with an embattled Boeing, embroiled in an all out all front war on the narrow body front, giving the launch a pass for the time being. The NMA, nonetheless, had sound conceptual contours & silhouette which could have potentially shaped up the NMA into a strong contender had Boeing acted swiftly & strongly on this giving it the due priority instead of its typical preoccupation with the wide bodies, may be just for a change!

The A321XLR program, thus, almost had a full afterburner take-off zooming to FL310 in a jiffy with a loaded cargo deck given the order book was standing at around 450 aircraft orders by the end of year 2019…with Boeing troops nowhere in sight as they were busy firefighting inside a hangar on fire…

Airbus, with the A321LR & XLR, thus, has successfully led a cavalry charge against Boeing by concentrating its forces to effectively exploit a clear gap right in the middle of Boeing's formation to run straight through the line potentially setting the stage for the launch of an encirclement maneuver of the entire narrow body landscape later; targeting the relatively & strategically weaker Boeing forces deployed & positioned with limited maneuverability; by unleashing a torrent of reserves & reinforcements.

Boeing's Counterattack Saga: a Tale of Missed Opportunities, Yet Again! – Y1 & NMA

Boeing's replacement plans for the 757 program starting in early 2000s and its subsequent counterattack plan to take on Airbus' aggressive crusade into the middle ground had centered on a weird concoction of strange, homebrewed acronyms, which have gone as: NMA (New Mid-Size Aircraft/Airplane), NSA (New Small Airplane) v1, NSA v2, NLT (New Light Twin), FSA (Future Small Airplane) & MOM (Middle of Market) etc. with none of these concepts actually materializing into any actionable, crystallized form with the Boeing furnace never able to reach the required temperature levels to be able to cast the die.

Boeing's strategy for narrow body replacement has straddled the thin red line demarcating the boundary between the traditional legion of mainstream single aisle aircrafts, led by the likes of 737, and the middle ground where the niche 757 traditionally ruled as the sole predator of the fully marked territory.

Boeing's replacement strategy had been yawing between Y1 & the NLT. The original replacement plans had been anchored on the Y1 leg of the Boeing's ambitious Yellowstone Project at the turn of the century with the Y1 designated as the composites based next generation successor of the 737 & 757 narrow bodies with fuselage stretching & minor modifications to create the two variants with significant commonalities and a common type rating.

This plan; anchored on a common narrow body clean sheet successor; which also was designated as the NSA & later evolved into the FSA, was one of the flight paths under consideration by Boeing while heading towards a successor. However, the Y1 was deferred in the 2000s decade at the concept stage itself with Boeing instead taking off on the 787 on its first 21st century mission to hunt down the A380 in a high stakes aerial dogfight. The NSA, too, was abruptly shot down prematurely with Boeing reactively forced into going for re-engining of the 737 into MAX for the need of a rapid counter fire to protect & hold the line as well as its competitive position in the almost bloody trench warfare with Airbus.

The other flight path which had been active on Boeing's routes map since 2011 had been the New Light Twin (NLT) which later gave way to the NMA and was intended primarily to check Airbus' growing ambitions & incursions in the middleweight class with the fielding of a new pugilist with proven antecedents as well as an impressive pedigree.

The NLT/NMA ambitious plan had a new, clean sheet twin aisle design at the core coupled with the operating economics typical of a traditional single aisle airplane to create the sustainable competitive edge. This plan entailed a bit of relatively higher degree of risk as it was going to be a niche airplane and numbers, in terms of size of the market opportunity, would matter much more than in the NSA plan, which had much more flexibility configured into the outline right from inception. Boeing's dilemma has remained around which horse to place the bet on amid a clearly surging Airbus stallion in the lead.

The NLT & NMA; since the onset of 2010s decade, have been Boeing's most serious efforts towards replacing 757 with a renewed Airbus charge, led by the new breed of voracious 'neo' predators eating rapidly into market shares, giving the much needed propulsion to Boeing's counter efforts.

This was actually a déjà vu situation & moment for Boeing which surprisingly had already faced a similar conundrum in the 1970s when Airbus had just appeared as a rookie on the radar. Boeing had been looking to replace the 727 tri-jet narrow body covering the mid-field while simultaneously it also had to defend against a tenacious onslaught from the larger Airbus A300/310 wide bodies.

After grappling with the dilemma for almost half a decade in the 1970s studying & exploring 7N7 & 7X7 concepts; considering all scenarios as well as possible solutions and ruling out a fuselage stretching based derivative approach and armed with newer generation turbofans & avionics; Boeing chose to go for an all out charge announcing the parallel development of 757 & 767, even conceptualized as almost fraternal twins, with the latter launched first in mid-1978 followed by the 757 in 1979.

The two aircrafts share a lot in common in terms of design philosophy, cockpits, avionics & aircraft production processes with the twins sharing a common type rating.

The commonality concept had originally been used by Boeing in the late 1950s with the creation of the successful 707 based aircraft family; comprising the smaller Boeing 720 and the militarized workhorses based on 707; with the 367-80 prototype being the common ancestor.

The adoption & pursuit of this parallel development approach brought & kicked-in additional financial savings & efficiencies through the development processes and transpired into significant operating advantages for the airlines as well who almost gave an instant thumbs up reception to the 757/767's overall value proposition.

The 1970s, however, was a different league & ballgame for Boeing, which by then had been positioned strategically as the undisputed market hegemon with the defense heavy McDonnell Douglas nibbling at Boeing's heels in the commercial market while the third vector, a primarily defense focused Lockheed Martin, exiting the commercial market in the early 1970s, the decade of the oil crisis.

Airbus; as the new European challenger; had just appeared on the aviation circuit with the pair of world's first twin-engine A300/310 wide bodies, with a no-nonsense workhorse approach being their sole piece de resistance.

The A300/310, with their clichéd design & staid appearance, must have come as a radical departure from the dead hand of the older quad & tri-jets glamorous era. The A300/310, as expected, did not create much noise at their EIS in mid-1974 as their take off rumble got more or less eclipsed by the sonic boom created by the world's first operational supersonic commercial jetliner, the rather superbly chic, Anglo-French Concorde, the almost Michelangelo of commercial aviation, which had its maiden flight in 1969 and entered service during the same decade, 2 years after the A300, in 1976.

The advent of the A300/310 had been overshadowed almost completely by the spectacle of the afterburner powered take-off of Concorde along with its technological advances like incorporation of an automatic flight control system (analog) & a light weight, aluminum alloy based airframe.

Concorde's 4 Rolls-Royce/Snecma built Olympus 593 engines actually were afterburning turbojets capable of producing 6,000 lbf[11] of additional thrust in wet mode with afterburners on, first ever for a commercial aircraft.

The Concorde had almost instantly usurped the bragging rights from other mere mortal commercial jetliners & hogged headline slots across most parts of the world (with its superpowers) along with scoops around Boeing's retaliatory fire plans with a SST (supersonic transport) of its own in the pipeline, which collectively dominated the aviation scene. The Concorde's high procurement & exorbitant operating costs as well as niche positioning was ensued by a limited production run of 20 aircrafts; thereby rendering it almost in the same unique, elite & rare class of the aviation's crème de la crème as the North Grumman Corporation built USAF's B-2 Spirit Stealth Bomber in the early 1990s with an equally short production run; and far away from being a threat of any kind to Boeing from even the farthest possible stretch of imagination. Both actually & deservingly have been the peerless ne plus ultra of commercial & military aviation respectively.

The Airbus A300, in fact, almost had to slog it out in the 1970s skies over grinding, back to back aerial sorties to prove itself and make its mark felt. Boeing, thus, in the 1970s had almost the entire midfield to itself along with the invaluable leisure of time to play around with and maneuver the skies as it wished.

The twenty first century commercial aviation, however, was a different league altogether with the rise of Airbus as an equal & formidable foe at the turn of the century with the market's evolution into a clear duopoly marked by an almost evenly split market shares. Boeing's long standing privileges of choosing market maneuvers & their timings unilaterally were thus curtailed to some extent with the addition of a competitive relativity vector into the equation.

This was much akin to the situation facing Roman Army legions & their commanders, who in their four centuries long protracted battle against the barbarians along Danube, came to realize after being humiliated by the barbarians in the Battle of Teutoburg Forest that the Barbarians across the Danube had become a force to reckon with and thus switched operating modes with the adoption of a strategic defensive posture.

This was marked by the treatment of limes/created fortifications along the Danube as the fixed, sacrosanct northern border of the empire with the focus from thereon remaining laser focused on tactically holding the line & protecting territory in the North instead of operating in the usual territorial expansion mode.

This kind of a submissive reaction was completely uncharacteristic of the usually trigger happy Romans as seen against the backdrop of their centuries long illustrious history of warfare. The Romans in fact, much earlier in the B.C. era; despite suffering a crushing defeat & massive losses against the Carthaginians in the battle of Cannae during the Second Punic War (with the victory scripted by Hannibal's classic pincer attack); had vowed to seek vengeance and they ultimately settled the score years later by obliterating Hannibal at the battle of Zama while also bringing the Carthaginians down on their knees forever.

Thus, this complete about turn in the Roman strategic posture to defensive against the barbarians, marked by a major deviation from their usual offense heavy operating mode, was what marked the beginning of the end for the formidable & legendary empire.

The history has somewhat been repeated by Boeing in its past decades of warfare against Airbus in the narrow body aircraft market with Boeing mostly staying on the defensive & reactive against the invader letting it become a formidable foe and whenever it tried to go on the offensive the half baked maneuvers backfired.

On the NSA; with Boeing (following the launch of A320neo by Airbus as the first mover) pushed to play on the defensive; almost had to abort the clean sheet based original mission profile at the planning stage itself and forced by competitive pressures & market forces to instead switch to the revised, re-engining mission profile for a shorter take-off run and a quicker time to market, that ultimately led to the somewhat hasty & reluctant launch of the 737 MAX. Same has been the story for Airbus which was almost unwillingly forced into taking a clean sheet approach on the A350XWB by the very same competitive & market forces.

The launch of the 737 MAX in August 2011 effectively meant dead end for the NSA pathway with the GPS coordinates reset & aligned along the NMA route. Boeing, looking to fill the huge void between the top end of the 737 MAX and the entry level variant of the 787, had initially in the 2000s dubbed the market development for pursuit of 757 replacement with a clean sheet aircraft design as premature as the order intake for 757 program towards the turn of the century had been abysmal across variants and the in-service fleet of 757s was relatively young given that the 757 order book really gained traction almost towards late 1980s thereby delaying the window of potential replacement demand's arrival to much later than usual.

Poked by the launch of the A321LR by Airbus in early 2015, Boeing had even studied & considered a quick fix with a re-engined 757 (dubbed the 757MAX) as a strategy option, however, the larger in size 757 (than a typical 737 kind narrow body), would have needed almost 30% more powerful engines than the existing LEAP-1A or PW1100G family engines built for the mainstream narrow bodies.

Additionally, any potential 757MAX would have needed a new, redesigned high aspect ratio wing apart from the requirement of a new engine capable of producing the niche thrust output class for the Maverick narrow body which couldn't have been met by an existing, off the shelf medium thrust engine.

A new engine derivative would have to be developed by either Pratt & Whitney and/or CFMI, both of which already had too much on their plates as P&W was busy tackling in-service issues of GTF engines while concurrently expanding the herd by developing newer variants across programs, in addition, to undertaking a rapid ramp up of production output. CFM International's industrial base, on the contrary, was deeply engrossed in its historic & simultaneous transition from CFM56 production ramp down to LEAP production ramp up.

Boeing's industrial base, too, was fully packed with an aggressive production schedule with the existing assembly lines running full throttle spewing out aircrafts to meet hectic delivery schedules and it would have been another high magnitude challenge of its own for Boeing to zero down on an existing facility & assembly line for the required additional capacity to produce the 757MAX.

Ultimately, the relatively larger size & niche positioning of the 757 ensured that the overall & relative math didn't work out with the re-engined 757 still falling behind the A321LR in relative operating economics. As with all things in life, even the Maverick 757's biggest strength thus had turned & played out to be its biggest weakness as well in an altered operating context!

The NMA was the only piece left on the table now for Boeing to solve the middle market puzzle & counter Airbus maneuvers. However, being a niche aircraft, NMA would have provided limited flexibility as well as maneuvering leverage to Boeing and the margin for error, thus, was going to be very limited and getting the numbers precisely right while taking the call was going to be critical for success riding on a potentially huge $10 billion to $15 billion bet.

This is the biggest concern & potential downside while going niche as one has to be precise and spot on without any room for errors or deviations and bigger the stakes bigger the risk. After all, an outlier is never mediocre!

United, looking for a replacement for its fleet of old 757s; again set up the stage for another battle by initiating consultations with both Airbus & Boeing for exploration of potential replacement options for its 757s in March 2015[12] at the ISTAT (International Society of Transport Aircraft Traders) conference. Boeing's Vice President of Marketing, Randy Tinseth, brandishing a broom handle at the same ISTAT conference venue in March

2015 retorted, "We're looking at the market, not an airplane we used to deliver. We're at that stage now where it's about collecting information. The launch of any new midsize airplane would likely have to wait as long as a decade, after Boeing starts delivering the 777X[12]."

This was logical & a no brainer as Boeing's R&D bucket had been already full with the ongoing development of the 737 MAX (since launch 2011) and it's yet to be reincarnated variants (-7.-8,-9,-10), the longest variant of the 787 family, the 787-10 apart from the development of 777X in two variants (-8 &-9) launched in late 2013, the re-engined version of Boeing's most successful wide body aircraft program, the original 777.

Additionally, the NMA would have needed a new, custom built powerplant with the specific thrust class which would also have added another few years to the timeline.

All of these collectively would have left Boeing with no margin for the concurrent pursuit of a new, clean sheet NMA at least through the end of the decade leaving Airbus in an enviable position, with the European plane maker more than happy to pick up the gauntlet and ride solo on the unleashed market bull-run ensuing the launch of the extended legged variants of the A321neo, whose order book had already crossed the 1,000 firm orders threshold by March 2016 as compared to 737 MAX-9's tally of just under 300 orders!

The NMA was going to be Boeing's only second new clean sheet aircraft program of the 21st century after the 787 and was quickly dubbed 797 by the ever so scoop hungry press folks & the industry enthusiasts.

Within months, at the Paris Air Show in 2015, Boeing's commercial airplane sales chief John Wojick confirmed that Boeing felt that the market was large enough and the company was conducting detailed study of the business case for the same[13]; after having brushed it aside only a year ago in

2014, courtesy the 'A321LR effect'.

However, Boeing, at this stage, still hadn't gone deep enough & was yet to figure out the specs, configuration, where to build it, build it solo or share costs with partners and whether to make it as a traditional single aisle (like the original 757) or go for a twin, make it out of composites or take the beaten metallic path.

Boeing even considered a choice between a longer 737 MAX stretch and a new scratch up NMA design. A MAX route would have encountered huge stumbling blocks on the engineering side given the limitation of the 737 airframe to sport larger, more powerful engines (to be able to match larger neos) and persisting & overcoming them by going around them would have led to the creation of an even disastrous sequel of the one already in the making, had Boeing chosen to go the longer MAX stretch way.

Boeing, thus, was literally at sixes & sevens, busy brainstorming with the top 757 operators pack led by United, American, Delta and Icelandair, some of whom had already started switching to Airbus stable having placed initial, launch orders for the A321LR, thereby, putting pressure on Boeing to pick up the pace on track to ensure that Airbus does not get away with an unassailable lead to conquest and its subsequent coronation as the new monarch of the commercial aviation's middle kingdom!

Boeing, in its initial market assessments, outlined the initial specs for a 757 sized clean sheet aircraft in 2016, to be developed in 2 variants, with a slightly longer range at 4,500 to 5,000nmi & a capacity of 200-250 passengers (with the larger variant covering the 280 seat mark) in a seven abreast cabin with an elliptical fuselage cross section[14].

The NMA, with a price tag in the range of $65 to $75 million, was going to be powered by a pair of new 40,000-45,000 lbf turbofans. Boeing's estimates pegged the market size for the NMA at around 4,000-5,000 aircrafts[12] over 20 years with an EIS target of towards the middle of 2020s, as outlined by Boeing CEO Dennis Muilenburg at an investor conference in February 2019 joined by Rolls Royce which concurred obsequiously with a similar view towards NMA.

The industry, led by GE Aviation & Pratt & Whitney, estimated the market to be around 2,000 to 2,500 aircrafts[15] at best which could have been a hurdle as the decision to plough billions into the development of a new engine program hinged very much on it.

Further, a potential, dual-sourced decision by Boeing riding out with two engine suppliers on the NMA would further complicate the matters and the engine manufacturers, thus, had been rooting for a sole source decision on engines.

The NMA, as the potentially first narrow body clean sheet program of the 21st century coming from the Airbus-Boeing duopoly, had already set off another ferocious battle of engines amongst the engine manufacturers looking for a share of the sizeable, long term growth pie.

The action packed, aerial combat thriller in the making was again going to have Pratt & Whitney, with its PW1100G GTF engine family, and the CFM International's LEAP engine family as the leading war birds engaged in high G aerial combat maneuvers with Rolls Royce jinking and eventually thrust vectoring its way out of the combat zone given the protracted EIS timelines for its latest, under-development Ultrafan engine.

Airbus; with its narrow body product portfolio strategy well in place and fully sorted out with the addition of the acquired A220 and the extension of the A321neo into LR & XLR variants; was cruising effortlessly hurtling along a high subsonic airspeed at FL350 with minimal drag, no signs of turbulence & well assisted on its course by a strong & vigorous market jet stream.

With the industry pressing Seattle to respond while Boeing contemplated hard, trying to play it safe in its trademark style, inching its way into the nebulous NMA territory; veteran Airbus sales commendatore, John Leahy; who, by now, had outlasted & outshone almost a dozen competing Boeing sales chiefs in the ring with his unmatched ringcraft demonstrated & showcased effectively in his over two decades long career with Airbus; took a dig at Boeing, throwing punches, in an interview over NMA as reported by Seattle Times.

He fired his precise sniper shots in his usual repartee: "The industry has no need for a new midmarket plane, since he is selling one: the A321neo[16]". He further added: "There's a need in Seattle, maybe. The A321 is clobbering them. We will have the luxury of sitting there, looking at what they do, and answering at a later date" with something better[16]".

Airbus take on NMA (as expounded by John Leahy) with a clean sheet approach was outright bearish with the company estimating the market size to be around 2,000 aircrafts which just did not provide a sufficient & justified business case to recoup the huge $15 billion investment required and to be able to book profits.

Leahy said. "If you bring out an all-new, clean-sheet airplane, you're taking a market there that is probably 2,000 airplanes. It's not 4,000, and you're spending $15bn to develop the airplane. You've got to amortize that and every airline is telling you $55m to $75m is all we're going to pay for this light twin. You've got the engine guys telling you the engines are going to be very expensive. The numbers just don't work[17]".

A stentorian Leahy further retorted post his retirement (at the end of 2018) taking a jab at Boeing: "If Boeing were to ask my opinion, which it hasn't, I would tell them I don't think there is a business case for that. If they want to build something, I would advise them to put their money into building a single-aisle airplane and then that could roll over to replace the 737 MAX in the middle of the next decade"[18].

Boeing had been looking for a price tag of around $100 million[20] for the NMA keeping a 25% margin for itself, as per Scott Hamilton of Leeham News. However, the 787-8, with an almost 40% longer range than the NMA, had already been available to the market at around $115 million (post discounts amounting up to almost 50% as a rule of thumb). Airbus has so far not publicly mentioned the list price of the A321XLR but taking the list price of A321neo as the base and factoring in the discounts usually offered to airlines and additional costs for the XLR's additional range, the number would come close to around $75 to $80 million. Boeing, thus, couldn't have effectively sold the NMA for more than the $75-$80 million market price ceiling and producing the NMA at $50 to $55 million a-piece to recoup the $15 billion initial investment would have been really tight.

Boeing's protracted dilemma on the NMA, which had become more of an inertia inducing bromide by now, dragged on with the NMA concept daydreaming & waiting patiently to be incarnated & designated as the 797 following official launch clearance as the MAX variants and the 787-10 completed their ground rolls, got airborne and disappeared into the clear blue skies.

By the end of 2018, the status quo prevailed with Airbus firmly entrenched in the driver's seat right in the middle of the market.

The Airbus constellation, however, was much less brighter than usual by the start of 2019 with three of its biggest & brightest stars fading away in quick succession with the veteran sales chief & COO-Customers, John Leahy being the first to depart the ship retiring in early 2018 with Toulouse bidding a special & somber adieu to its 'Il Commandant', Leahy, who left really big pair of shoes behind to be filled.

His legacy: the maverick playbook & his repartees and it would be great to hear straight from the horse's mouth about his adventures across the race tracks & courses, in what would be an invaluable & almost canonical work for commercial aviation; somewhere down the road!

Leahy's departure was almost concurrent with the COO, Fabrice Brégier's exit, who pulled the lever to eject over prevailing incessant, internal power struggles to seek greener pastures outside.

This was followed by CEO Tom Enders departure in early 2019, whose term was not extended by the board. The news of Leahy's retirement must have come as music to Boeing's ears as their bogeyman, having stayed on top of the company hit list for decades, was finally gone. Seattle must have exclaimed, Good riddance! The development paved way for the techie, Guillaume Faury, to smoothly hover down & land into the CEO's office in his chic & sleek H160.

At the onset of 2019, on the other side of the Atlantic, the Boeing camp had still been grappling with the NMA quandary, which had been lingering on in a year that was going to be a horrible one for Boeing and one of the worst in decades, with its most dreaded nightmares on the verge of coming to life to haunt the 'Spirit of Renton' and an equally spirited Boeing company!

"Strategy without Tactics is the Slowest Route to Victory. Tactics without Strategy is the Noise before Defeat "– Sun Tzu, The Art of War

Chapter - 11

'CRASH COURSE' FOR THE 'REINCARNATED WARRIOR' & A 'SHIP ON FIRE'

"You cannot escape the responsibility of tomorrow by evading it today". –

Abraham Lincoln

I t was 08:38 AM local time (05:38 UTC) on a bright spring morning on a Sunday in early March 2019 in Addis Ababa, Ethiopia. Amid the clear weather conditions and with good visibility prevailing, a new 737 MAX-8 aircraft belonging to the Ethiopian Airlines, delivered only a few months back in November 2018[1], was ready for the take-off roll on the runway 07R. It was the Flight 302 (ET-AVJ) scheduled to fly from Addis Ababa to Nairobi, Kenya with an experienced crew in charge in the cockpit, led by the Captain who had over 8000+ cumulative flying hours to his credit, of which 1500 hours had been on the Boeing 737[1]. The Ethiopian Flight 302 pilots, with 149 passengers (originating from 35 countries) onboard along with a crew of 8, completed the take-off run and the airplane lifted off the runway.

However, much to the pilot's appall the new airplane (having logged barely1330.3 hours and a total of 382 cycles so far[1]) struggled with ascend and within a minute of taking off, a flight control problem was received by the control tower as reported by the onboard first officer. Within 2 minutes of the take off the sinister MCAS system onboard the MAX got activated pitching the plane's nose down & plunging the plane downwards towards the ground.

Within 3 minutes into the flight, the aircraft kept losing altitude and with engines on full take-off power it had also been accelerating rapidly. The unsuspecting pilots; completely unaware of the presence of something called MCAS, on a new aircraft which was supposed to have similar manual, non-fly by wire flight controls just like all other previous members of the 737 aircraft family; had been frantically all over the cockpit controls trying to figure out a way to somehow stabilize the aircraft. With the aircraft unable to maintain flight path, the pilots sent out the SOS and requested for a return to the airport for emergency landing which was granted.

However, by 08:44 AM (05;44 UTC), within just six minutes of the take-off, it was all over for the beleaguered aircraft which lost all contact with the air traffic control tower, went off the radar screens and crashed in a farm field 28NM South East of Addis Ababa with no survivors[1]. This had been one of the deadliest airplane crashes in the history of Ethiopian aviation since the hijacking and subsequent fatal crash of the Ethiopian Airlines Flight 961 in the mid-1990s.

An extract from the post accident preliminary investigation report[1], prepared by the Aircraft Accident Investigation Bureau (AIB) of Ethiopia, had the AoA & the MCAS pinned down as suspects:

"The takeoff roll appeared normal, including normal values of left and right angle-of-attack (AOA). Shortly after liftoff, the value of the left angle of attack sensor deviated from the right one and reached 74.5 degrees while the right angle of attack sensor value was 15.3 degrees; then after; the stick shaker activated and remained active until near the end of the flight. After autopilot engagement, there were small amplitude roll oscillations accompanied by lateral acceleration, rudder oscillations and slight heading changes; these oscillations also continued after the autopilot disengaged. After the autopilot disengaged, the DFDR recorded an automatic aircraft nose down (AND) trim command four times without pilot's input. As a result, three motions of the stabilizer trim were recorded. The FDR data also indicated that the crew utilized the electric manual trim to counter the automatic AND input. The crew performed runaway stabilizer checklist and put the stab trim cutout switch to cutout position and confirmed that the manual trim operation was not working[1]".

This deadly crash of the Ethiopian Airlines Flight 302 was the second one for the 737 MAX-8 within a 6 months span and was preceded by an eerily similar, another crash of a brand new 737 MAX-8, belonging to the Indonesian Lion Air, in October 2018. Lion Air Flight JT610, flying on a domestic route carrying 189 passengers, commanded by a seasoned captain with over 6,000+ flight hours[2], had nosedived into the Java Sea on October 29, 2018 barely 13 minutes after take-off in a strikingly similar manner (to the Ethiopian crash) leaving all on board the aircraft dead and a trail of unanswered questions pertaining to crashing of a brand new 737 MAX-8 which had just been delivered two months back in August 2018 and had clocked a mere 800 hours[2] and had its sole Angle of Attack sensor had been repaired barely a day before the crash due to inaccurate readings.

The Indonesian crash, however, had been somewhat eclipsed by the country's relatively shady aviation safety track record; with Indonesian carriers also getting operationally banned by the European Union (EU) in

2007 in its Air Safety List over their inability to comply with international safety standards, with the ban having just been lifted in June 2018 by the EU[3]. The latest 737 MAX-8 crash thus, had, been treated initially, as another & latest addition to the country's existing list of air crashes.

Preliminary investigations into the Lion Air crash, however, revealed that the aircraft's previous flight, prior to the crash, too had been a harrowing one and had a narrow escape. However, the latest one did not have luck on its side and could not overcome the in-built monster, the MCAS.

Harrowing accounts of those on board the aircraft, including, the crew as well as the passengers, as the pilots wrested for control of the aircraft, on both the flights raised potential red flags over MAX's flight control issues, Angle of Attack sensor deviations, instrument failures & abrupt nose pitch downs, all emanating from MCAS, which had no mention in the aircraft's flight manual as well as the pilot's computer based training (CBT) refresher created for the MAX by Boeing.

The dynamics of the aircraft's final minutes prior to the fatal crash had been characterized by the very same 'rapid acceleration' & 'high descent rate' within minutes of take-off, as tracked by the flight tracking service, FlightRadar 24. Consequently, warning & training advisories were issued by the FAA as well as Boeing for global operators of the 737 MAX series aircrafts pertaining to the MCAS and the abrupt nose dives caused by it on the aircrafts with Boeing publicly revealing & acknowledging for the first time the presence of the automated flight control system, MCAS, on the 737 MAX in November 2018.

Within a week of the Lion Air crash, Boeing sent a service bulletin to airlines followed by an airworthiness directive from the FAA specifying a flight recovery procedure[4]. But the question is was that enough? It was not going to be as these recommended recovery procedures could not save the Ethiopian Airlines 737 MAX-8 from crashing 6 months later.

Following the second air crash involving the Ethiopian Airlines Flight 302, the entire global fleet of in-service 737 MAX-8 aircrafts was grounded for safety reasons given the similarities between the crashes starting with Ethiopia, followed by China with most of the aviation regulators following suit and the FAA being the last one to issue grounding directive on the U.S. based MAX fleet with a total of around 400 737 MAX-8 aircrafts, delivered by Boeing, operating across the globe grounded by late March 2019. The role of MCAS as the common link in both the crashes and fleet grounding was followed by a series of serious whip cracking on Boeing from all factions & directions demanding action as well as accountability.

There were two very similar, fatal air crashes involving two brand new aircrafts crashing within a span of 6 months in clear weather conditions, with a weird MCAS feature factoring twice into the equation, which had been a clear contrast to & a major departure from the usual air crash news pieces which typically entail old aircrafts operating in rough weather conditions.

This skewed the needle of suspicion in the regulatory & public compasses further towards Boeing for potential design & safety issues, especially, given the Ethiopian Airlines' watertight safety record. This was going to be the watershed moment for Boeing and a beginning of the downhill journey of the aviation giant from the top of the pike. Multiple investigations into the FAA's certification process on the 737 MAX were initiated led by the U.S. Congress, Department of Transportation and the FBI followed by the minute & close parsing out of the aircraft development

& certification processes.

Spurred by the twin, freak crash accidents, aerospace reporting veterans covering the sector; who had already smelled a rat; spearheaded by a feisty Dominic Gates & the team of The Seattle Times reporters, dug their way deeper into the trenches looking for traces & trails unearthing evidences lying buried deep beneath the surface, unraveling a sinister plot of major compromises & lapses on safety front green lighted by a hubristic Boeing leadership, who had been operating in the overdrive mode to catch up on Airbus market lead in the narrow body aircraft market somehow or anyhow, even at the cost of safety.

The audacious plot was uncovered by the Seattle Times journalists (for which they deservingly have won the Pulitzer Prize for 2020) along with whistleblowers. The plot was marked by a serious lack of regulatory controls & deficiencies coupled with unfathomable lapses & gross oversights in the certification process on 737 MAX on part of the FAA, which in turn had delegated most key parts of the certification processes to Boeing's technical employees who did the work internally toeing the line laid out by their numbers focused ringmasters who kept everything else at bay.

The plot also exposed Boeing's inexorable pursuit of profits even at the cost of public safety, which had clearly been put at stake. For the uninitiated, those getting out of their caves after a long hibernation or those returning to Earth from deep outer space or those looking to further deep dive may explore the well chronicled account of the same by checking out the coverage by the Seattle Times[7] and The Air Current[8].

---

[7] https://www.seattletimes.com/business/boeing-737-max-crisis-2019-news-coverage/

The final investigation report, released in October 2019 by the Indonesia's National Transportation Safety Committee (KNKT) prepared in assistance with the U.S. National Transportation Safety Board (NTSB), apportioned the crash blame to "a combination of an improperly aligned angle of attack (AOA) sensor, lack of pilot reporting and training as well as a breakdown in safety oversight of certification and design flaws shared between Boeing and the FAA within the aircraft's maneuvering characteristics augmentation system (MCAS) system[5]".

The KNKT's final accident investigation report on the crash also zoomed in and had its stranglehold on the MCAS going straight out with the charge, "the expanded capabilities of the system to even low Mach number situations and increase to maximum command limit of 2.5 degrees of stabilizer movement led to occurrence of unintended MCAS-controlled stabilizer movements. During the 737 MAX's process of certification, these MCAS triggered unintended stabilizer movements were considered to be major & significant, which, however, with Boeing's contention that these could be offset & countered manually through stabilizer trim and cutout actions, prevented any further analysis of its failure conditions[5]".

The point is that if Boeing omitted even the mere mention of the existence of the MCAS from the flight manual and pilot training material, how were the pilots supposed to counter the effect manually when they were not even aware of the cause of the effect in the first place.

The report by KNKT also pointed to the malfunctioning of the MCAS system based on a single Angle of Attack (AOA) sensor (out of the 2 sensors available on the 737) as compared against 3 AOA sensors used by

---

[8] https://theaircurrent.com/aviation-safety/vestigal-design-issue-clouds-737-max-crash-investigations/

the A320 aircraft family's on-board computers. Basing the MCAS' all sweeping, demonic powers on a single AOA sensor configured the ailing system precariously positioned on a single point of failure.

This also led to jettisoning of the principle of built-in system redundancies which has been the guiding beacon, principle & virtually the bedrock of modern age aviation safety. However, in this case it had simply & outrageously been overruled & brushed aside on the MAX.

The Ethiopian aviation regulator, ECAA, established, based on the investigations, that the flight crew had attempted the recovery procedure outlined by Boeing in the service bulletin issued post Lion Air crash and squared the blame on the aircraft's inherent design flaws echoing the Indonesian regulators view. The U.S. NTSB said "The assumptions that Boeing used in its functional hazard assessment of the un-commanded MCAS function for the 737 MAX did not adequately consider and account for the impact that multiple flight deck alerts and indications could have on pilots' responses to the hazard. Boeing failed to assess the consequences of MCAS failure and made erroneous assumptions about flight crew response[6]".

A report by the U.S. Department of Transportation's Office of Inspector General (IG), released in June 2019 said "Boeing deliberately misrepresented MCAS to avoid scrutiny[7]" as the MCAS had been presented to the FAA by Boeing as a modification of the aircraft's existing speed trim system with limited usage & application.

The report also noted that an FAA assessment, made as early as in December 2018 post the Lion Air crash, had already assessed & projected that if the MCAS software was not fixed it could lead to around 15 crashes over the entire life span of 737 MAX[7] but this was not revealed at the time by the U.S. aviation regulator.

A Congressional committee probing Boeing over MAX crashes for over a year criticized Boeing's "culture of concealment" [8] in its preliminary summary of the report in March 2020 after having held 5 hearings (which even grilled the Boeing CEO Dennis Muilenburg, who testified in October 2019) and having uncovered & devoured tons of Boeing's internal documents clearly outlining safety concerns of Boeing employees over MAX[8]. Boeing's own employees had been outright nasty about the shoddy MAX development & certification process, as revealed by the exchange of internal communication & documents through e-mails & messages, with an outrageous employee comment, published in a New York Times article in early 2020, saying that "this plane is designed by clowns, who are in turn supervised by monkeys[9]". Another pilot working on 737 MAX development, had told a colleague that "he was experiencing trouble controlling the Max in a flight simulator and believed that he had misled the FAA.[9]"

A whistleblower; a Boeing engineer working on flight-deck systems during the 737 MAX development; wrote to a U.S. Senate committee after the Ethiopian crash over the outrageous MAX development process highlighting serious design flaws in the airplane, following an internal ethics complaint filed earlier,[10] as revealed by Seattle Times. This engineer in 2014 had recommended incorporation of a new synthetic airspeed system (already available on the 787) to upgrade the MAX's original flight controls for enhanced safety.

However, his recommendations were turned down by Boeing management over budget & time constraints along with additional pilot training requirements it would have necessitated. He wrote in his letter:

"These flaws were known to Boeing as it worked with the FAA to certify the 737 MAX, and awareness of this was creatively hidden or outright withheld from regulators[10]".

The Congressional Committee report further said: "Boeing had undue influence over the Federal Aviation Administration, and FAA managers rejected safety concerns raised by their own technical experts[8]". A criminal investigation had been initiated by the Justice Department as well. All hell had broken loose on Boeing which by now was being pounded & soon going to be ravaged by an almost Category 5 hurricane under the floodlight & scrutiny of all & sundry.

Captain Chesley Sullenberger; better known as Sully, who had achieved the superhumanly feat of safely landing a U.S. Airways A320 post twin engine damage & subsequent failure due to bird strikes on the Hudson River over a decade back; simulated the crash scenarios on a MAX simulator and he, too, ruled unequivocally in favor of the pilots.

He told a congressional subcommittee, reflecting on the incident in light of his past & simulator experiences, as reported by NPR[12],"I can tell you firsthand that the startle factor is real and it's huge. It absolutely interferes with one's ability to quickly analyze the crisis and take corrective action. Even knowing what was going to happen, I could see how crews would have run out of time and altitude before they could have solved the problems"[12]. He further added, "We should all want pilots to experience these challenging situations for the first time in a simulator, not in flight with passengers and crew on board"[12].

In fact, Boeing's test pilots had already encountered & foreseen[13] the potentially catastrophic outcomes of a delayed reaction time of beyond 10 seconds on the flight simulators during MAX development way back in

2012; as per the detailed, thorough & comprehensive final investigation report of the House Transportation Committee released in mid-September 2020.

The report also found evidences of how Boeing's head of MAX program, Keith Leverkuhn, authorized MCAS activation at low speeds in March 2016 (without being aware of its expanded scope of operations & implications, as found by investigators) endowing it with sweeping powers, thereby, virtually & effectively turning it into nothing short of an onboard 'Terminator'. Simultaneously, Boeing's chief technical pilot, Mark Forkner, got the FAA to delete even the mention of the term MCAS from the pilot flight manuals in March 2016 itself with FAA officials kept in dark about the expanded functional profile of the system, as per the report[13], which further uncovered that Boeing, in fact, awarded Forkner & his technical pilots team in September 2016 for their contribution towards development of pilot training and getting the MAX certified without necessitating additional simulator training for the pilots!

The Committee chairman, Peter DeFazio, told the reporters, "Boeing presented a massive amount of data to the FAA but they never highlighted MCAS as a novel or new system, in fact they had an internal meeting in 2013 where they agreed they'd never talk about MCAS outside of Boeing" [14].

The Committee also lambasted the FAA chief for his unabashed denial of any irregularities with the MAX certification process and terming it as 'compliant' with Boeing toeing the same line as well. DeFazio has derided it as "mind-boggling" [13], [14] adding further that "Both the FAA and Boeing came to the conclusion that the certification of the max that killed 346 people was 'compliant' but the problem was it was compliant and people died" [13], [14]. The committee plans to bring in further legislation aimed at plugging the existing gaps and to ensure stringent processes for airplane certifications going forward.

Around 346 people had succumbed to and had become the unsuspecting martyrs who laid their lives to the avaricious, all subversive & nefarious mechanisms of the cronies of capitalism. This was the worst of the black ops missions ever to be green lighted by the dingy corners of Boeing's leadership barracks resulting in huge blowbacks which ultimately forced the company to reap the resulting, violent whirlwind.

The lineage of this sinister mission profile, however, clearly had the typical characteristics & radar signature which could easily be traced back to the 'old masters of creating cutting corners based disasters', McDonnell Douglas, with a comprehensive log book maintained & published by the NTSB and their regulatory counterparts across the world capturing & covering their diverse array of in-action aerial antics in a foregone era!

The 737 MAX-8 disaster had uncanny & strikingly eerie similarities and very strong parallels with the DC-10 debacle of the 1970s with the latter having earned a unique, highly dangerous reputation for itself well reflected by the plethora of infamous nicknames conferred upon it, which included, 'Death Contraption', 'Death Cruiser' and 'Daily Crash' etc. with the plane even fully capable of taking on military jets with its spectacular ability of seamlessly carrying out barrel rolls and flying inverted having demonstrated it effectively on multiple occasions. The four deadliest air crashes involving the DC-10 collectively accounted for around 1,000 precious lives[11].

The DC-10, however, got away with McDonnell Douglas' heft and unique modus operandi centered on crony capitalism augmented by the all pervasive, below the radar sinister networks of alliances & nexuses.

However, in the 21st century, dominated by a trigger happy media & social media, has brought in unprecedented levels of transparency worldwide which couldn't have kept the systemic design flaws & shortcomings of the MAX hidden for long.

Boeing, thus, had messed up big time on the MAX in its high stakes race to take on a renewed Airbus charge in the narrow body segment. As the traditional iron triangle of the model of constraints says, 'you cannot make something cheap, fast & good at the same time' and Boeing had crossed the sacrosanct line here trying to get all the three in one go. The iron triangle though has been defeated to some extent by digital age disruptive technologies, like 3D printing, but they were not enough to save Boeing's self created disgrace on the MAX front which was surely headed for the Aviation's 'Hall of Infamy'!

The MAX crisis had come to hit Boeing almost like raging shock waves & electromagnetic pulse following a nuclear blast and once the layers of the plot started unfolding the fallout & the radius of impact area started expanding rapidly. This was especially since Boeing had initially refuted any design flaws or faults with the 737 MAX until the evidences proved it otherwise. The loved ones & families of the 346 air crash victims pursued legal actions against Boeing seeking damages. The grounding of the 737 MAX fleet globally just prior to the summer peak season caused chaos & mayhem across airlines as their fleet planning & capacity had been severely impaired & crippled with all 737 operators, like Southwest & Ryanair, being the worst impacted of the lot.

These airlines had to revamp & alter their original, detailed flight schedules entirely at the eleventh hour amid a serious capacity crisis triggered by the grounding of their workhorses. The airlines sought millions of dollars in damages from Boeing for not being able to use the grounded planes.

With aircraft deliveries to airlines & lessors getting deferred and the sizeable order book for the 737 MAX almost going into limbo; Boeing's revenues & incoming cash flows were impacted directly. Cancellations of the already placed orders for the 737 MAX were also in the plot which worsened the situation for Boeing as well as the entire industry value chain, including, key suppliers on the 737 MAX program like Spirit Aerosystems which makes fuselages for the 737 aircrafts.

Further, with flyers and the general public showing utter lack of confidence in boarding the MAX even post its re-certification and the Boeing's battered stock price virtually in a tail spin mode; the worst ever disaster in Boeing's over a century old history had brought to the company's shores, nothing short of a crisis of the worst order.

Boeing's head of commercial airplanes division, Kevin McAllister, was the first head to roll within Boeing in October 2019, as the fallout of the MAX crisis started proliferating, with the action meant to signal to the market, public & regulators that the company was taking accountability for the disaster.

This was followed by the Boeing board's decision to fire the CEO Dennis A. Muilenburg in December 2019; who had been facing the large calibre regulatory cannons as the Captain of the ship which had blown up while being on a buccaneerish misguided escapade; a move (referring to firing of the CEO) the board said was 'essential to restore confidence' in the company, signing off reactively on Boeing's worst year ever in the company's long & illustrious history!

Muilenburg's Cameo in the Boeing Cockpit - Forced to Eject, Overpowered by the Onboard MCAS

Apart from McNerney, the other key face to star in lead role in the Boeing's commercial aviation motion picture in the 21st century has been his ultimate successor, Dennis Muilenburg. Muilenburg's profile has had many striking similarities with Philip Condit with both being hardcore engineers at the core and true blue Boeing bred stallions who started right at the base and made their way up through the ranks over decades growing within Boeing to become the Chairman & the CEO ultimately.

However, the only difference between them being that while Condit spent most of his early career in the commercial side of the Boeing's business while Muilenburg slogged it out mostly on the other side of the fence, in the defense business, especially during the second half of his career following a mixed stint across both sides in the earlier half. Muilenburg had been engaged by Boeing over multiple, high stakes defense programs in engineering as well as management positions with the most important ones being the X-32, unveiled for the JSF program (which went head to head against the Lockheed Martin's contender) apart from the YAL-1 747 Airborne Laser, the High Speed Civil Transport, Condor unmanned reconnaissance aircraft and as a program manager on the U.S. Army's futuristic Future Combat Systems (FCS) program.

Unfortunately, Muilenburg could never lead Boeing across the finish line as the winning horse on any of these defense programs as most of these exotic programs never saw the light of the day in flesh & bone amid prevailing budget realities while the high stakes X-32 was eventually lost by Boeing to Lockheed Martin.

Muilenburg ultimately was appointed as the President and CEO of Boeing Integrated Defense Systems (now BDS) in 2009 and he served there till his eventual appointment as the Boeing CEO in July 2015. It was under his reign at BDS that Boeing managed to lay its hands ultimately on the KC-X program. The keel for his appointment for the top job, however, had been laid in December 2013 with his appointment as Boeing's President as well as COO and as a potential subservient sidekick for McNerney.

Muilenberg learnt the ropes from him as it was going to be his first shot at the helm and as a perfect protégé vowed upon his coronation to continue with the same game plan crafted originally by McNerney going forward with minimal strategy deviations and avoiding any radical departures. At the time of 737 MAX launch decision in 2011, Muilenberg had been at the reins of Boeing's defense business as the President and CEO and seemingly did not have much direct bearing on the ultimate decision to re-engine the 737.

However, the 737 MAX's botched up certification process and the covert insertion of the MCAS piece by borrowing it from the KC-46A Pegasus and its subsequent mindless implementation led by its configuration & ultimate installation with full & absolute authority over pilots in the looping pattern surely were undercover missions duly authorized by his command.

Further, his tendency to use the available lines of communications & channels sparingly coupled with limited ability to effectively engage stakeholders & manage diplomatic ties were subjected to a severe stress test and a massive public fire trial later only to fall short and ultimately led to the ejection of another Boeing Chairman and CEO from the cockpit for the third time in the first two decades of the 21st century, following Philip Condit & Harry C. Stonecipher's departure earlier.

Muilenberg, in fact, ended up living the decisions made by his predecessor (McNerney) and consummated by him in a long cycle business which along with his atrocious handling of the post MAX crisis fallout, almost a PR disaster, made things worse.

This inability to show 'courage under fire' guided by truth, transparency & humility and to effectively transform it into almost what could have been 'Boeing's finest hour' could be traced & attributable to his relative inexperience at the helm coupled with his own personal shortcomings & persona for leading a ship in existential crisis emanating mostly from the proclivity to follow the GE-derived, Boeing's typical & comprehensive internal leadership textbook to the hilt with the SOP for dealing with the out of the blue, existential crisis scenarios just not there in the internally revered tome!

Muilenburg's Legacy

Muilenburg's legacy could best be summed up by the renaissance of Boeing's defense business under his reign with crucial wins over few key, long term franchise programs geared towards the future led by the USAF's T-X Trainer program with SAAB, MQ-25 Stingray unmanned aerial refueling aircraft for the U.S. Navy's CBARS/UCLASS programs, USAF's UH-1N replacement program with the MH-139 and the latest one (after his departure) being the selection of the F-15ES in 2020 by the USAF positing BDS on a strong footing for the long term.

His departure paved way for the implementation or rather repetition of the Boeing's board's time tested firefighting strategy of bringing in an old hand from GE's Aviation business already serving on Boeing's board. The strategy has been now repeated almost thrice by the board in the first two decades of the 21st century with Stonecipher, McNerney and now David Calhoun with all of them brought in to steer an imperiled ship back on-course.

Boeing's Chairmanship, over the past two & a half decades, thus, has oscillated repeatedly between thoroughbred, scratch-up Boeing engineers growing through the ranks & going on to summit the peak on one end (Philip Condit & Dennis Muilenburg) and the triad of GE Aviation's old hands on the other.

David Calhoun, the board's latest acquisition, too, has been thoroughly GE schooled having joined GE as one of the commanders of the Jack Welch's leadership juggernaut of the 1980s era going on to lead GE's Aviation business having steered the ship through the post 9/11 industry mayhem.

He has a huge responsibility now to see Boeing through the turbulence while carrying out a major overhaul under the hood to be followed by sorting out the product strategy mess and setting the company on course effectively for future on the most appropriate flight path aligned with the post-pandemic battered aviation landscape. Hopefully, he will bring an original game plan as well as flight plan!

737 MAX – The Boomerang Effect on Boeing

With the 737 MAX grounding, Boeing's commercial airplane deliveries for 2019 also took a nose dive to almost the lowest level in over a decade at 380 wide body airplanes, with 127 737 MAX deliveries for 2019 as against 580 for 2018, along with a deluge of cancellations on the 737 program (183 for 2019) translating into a net order intake of a mere 54 aircrafts for 2019, as per Boeing's Annual Report for 2019. Boeing's commercial airplane segment's top line shrunk almost by half for 2019 (down to $32 billion[12] from $57 billion in 2018) with the MAX deliveries halted since March 13, 2019 and over $8 billion in compensation paid to airline customers. The commercial airplane division booked a net loss of $6.6 billion for 2019[12].

Boeing also had reduced the production rate on the 737 program to 42 aircrafts a month from mid-April 2019 onwards, down from 52 earlier. This was in stark contrast to the originally planned increase to 57 per month. With almost an inventory of 450 737 MAX aircrafts parked around Boeing Field waiting to be delivered by the end of the year, the production had to be halted temporarily in January 2020 adding another $4 billion[12] financial blow in shutting down and restarting production.

Boeing also anticipated almost doubling of the overall cost of production of 737 MAX program to $6.3 billion in the post grounding scenario, up from $3.6 billion earlier[12]. The tally of order cancellations had swelled up to 600 aircrafts by June 2020 exacerbated further by the outbreak of COVID-19 crisis in early 2020.

Further, Boeing's order book for the 737 MAX program has depleted by around 1,000 aircraft units through the course of 2020 owing to order cancellations & removal of very weak orders; that are unlikely to convert to confirmed orders; from the order book, in line with the ASC 606 regulatory mandate. The net financial impact on Boeing, emanating from the MAX crisis by January 2020, had been assessed at a colossal $18.6 billion, as per the company's Q4 financial results.

Further, a quick look at Boeing's balance sheet shows a rapidly spiraling up concern from a long term perspective which is the long term debt levels. Boeing's long term debt for the year 2018, before MAX crisis, stood at $10.6 billion as per the 10K for 2018. This number had doubled within a year's span to almost $20 billion by the end of 2019, as per Boeing's 10K for the year 2019.

Also, with the 737 MAX's continued grounding, coupled with the COVID-19 outbreak in early 2020, Boeing's long term debt has tripled, having ballooned within a span of 9 months from $20 billion at the end of year 2019 to $61 billion by September 30, 2020, as per the 10Q filed for Q3 2020 with Boeing having issued fresh debt worth $42 billion so far in 2020.

Boeing's net loss for the nine months of 2020 before taxes stands at $5.8 billion with the commercial aviation business booking a net loss of $6.2 billion for the first three quarters of 2020. Further, Boeing's operations have used up $14.4 billion of cash for the nine months of 2020 with inventories, led by the undelivered 737 MAX units numbered at almost 450 aircrafts, accounting for $9.65 billion.

Boeing, thus, has burnt almost $19 billion of cash on inventories/undelivered airplanes alone in the 18 months period spanning March 2019 to September 2020 while having also created a mountain of debt for itself which would need to be scaled over long term. Boeing has assessed the overall damage to it from the 20 month long grounding at around $20 billion (by the end of Q3 2020) excluding the civil lawsuits over damages payable to the families of crash victims.

Further, with COVID-19 decimating commercial aviation in 2020, Boeing announced in response in April 2020 that it will downsize significantly and will lay off almost 10% of its 160,000 strong workforce by the end of 2021 translating into around 16,000 employees.

In October 2020, as part of the 10Q filing for Q3, Boeing announced further restructuring with another 7,000 jobs to be put on the chopping block owing to the double whammy of pandemic's aftermath and continued

737 MAX order cancellation spree. In a nutshell, Boeing clearly is struggling and has a daunting challenge ahead of itself over near term.

Fitch has already downgraded[15] Boeing's Long-Term Issuer Default Rating (IDR) and senior unsecured debt to BBB- from BBB in late October 2020 amid rising debt levels with a negative outlook given the pandemic's impact on commercial aviation coupled with MAX grounding.

Fitch projects[15] that Boeing will need at least 2 years to get its financials back on track and to the BBB level which is likely only by the end of 2022. The free cash flow for Boeing is also likely to be in the red zone at least through the first half of 2021, impacted by low-rate production, with the 737 MAX inventory also likely to be cleared only by 2022, as per Fitch[15] , in consonance with Boeing, which has opined the same.

The original contest for market shares & dominance, thus, turned into a storm of tragedy and a disaster with the mindless & relentless pursuit of economic profits thereby exposing the dark and the evil side of capitalism at its worst and its sordid underbelly. Boeing's reactive strategic orientation & operating mode on the 737 program for decades followed by resorting to cheat codes when it ran out of options had caused the worst ever crisis in the company's history.

The amount which Boeing has ended up coughing in compensations is almost 30% more than the $15 billion originally required for building a scratch up successor in time for the 737NG program.

In a sudden & dramatic turnaround in the overall scheme of things, the very company that had been the epitome of Aviation safety had its name inscribed in bold letters on the very epitaph of Aviation safety!

The road to recovery from the disaster is going to be an excruciating uphill climb with MAX becoming one of the most scrutinized aircraft programs in aviation history. The aircraft has been grounded for over a year and 6 months now and on July 1st the FAA decreed start of re-certification flights for the MAX.

By the end of October 2020, MAX had completed almost 3,000 flight hours of re-certification flights with the aircraft even commanded by the FAA chief, Steve Dickson, himself who was at the controls as part of an evaluation flight. Boeing engineers have now reconfigured the MCAS and have based it's activation on two AoA sensors while also wresting away most of the powers resting with the MCAS; especially the repeated, nose-down movement of the aircraft in a looping pattern which will now take place only once with the pilots now vested with the powers to fully overrule the system; by bringing it under the jurisdiction of the onboard flight computers.

However, with EASA and Transport Canada demanding major overhaul & augmenting of the current safety measures over the course of time; with EASA mandating Boeing to add a third synthetic AoA sensor and Transport Canada requiring a system to switch off the stick shaker alert system; it really has been a series of long & tedious efforts on Boeing's part to get the MAX off the runway and keep it airborne post comprehensive software updates & overhaul on its revenant amid strong crosswinds blowing.

Nonetheless, the MAX is likely to be in the air by the end of year 2020 or early 2021 with the FAA rescinding its March 2019 grounding order & lifting it on November 18, 2020 and thus effectively clearing the way for the 737MAX to get airborne again while also signaling the beginning of the end of Boeing's 20 month long nightmare.

American has already announced deployment plans for its fleet of 24

737 MAX aircrafts scheduled for late December 2020[16] which is going to be preceded by MAX flights without passengers. Boeing engineers will now be operating in the overdrive mode to get the grounded MAXs ready to fly while preparing the stored ones ready for deliveries post incorporation of requisite software updates & additional changes as mandated.

Boeing also needs to get the new pilot training program, developed for the 737 MAX for each U.S. based carrier operating the airplane, approved by the FAA before roll out, in addition, to getting the airplanes, produced since the grounding order was issued, recertified by the FAA.

However, American also plans[16] to explicitly inform the customers that they are booking on the 737 MAX aircraft! Thus, getting the 737MAX back in the air is just the first hurdle but regaining the flying public's trust and recuperating from the immense reputational damage will be a big ask & is a real litmus test that lies ahead for Boeing going forward.

Boeing, in the meanwhile, has been busy rebuilding the depleting order book for the 737 MAX program by booking new orders, since the lifting of the grounding order by the FAA, while offering significant discounts to airlines. Leading European LCC, Ryanair; booked orders for 75 new 737 MAX 8 aircrafts on December 3, 2020 taking its order book for the 737 MAX aircrafts from 135 to 210 worth almost $22 billion. Alaska Airlines, too, has increased its 737 MAX airplanes on order with Boeing with the announcement of the decision to buy an additional 23 737 MAX 9 airplanes on December 22, 2020 along with options to buy another 52 MAX aircrafts, thereby, becoming the first U.S. Carrier to place orders for MAX airplanes since FAA's recertification of the program.

Another key development, as a fallout of the MAX crisis, has been the passing of the Aircraft Certification Reform and Accountability Act (H.R.

8408)[17] by the U.S. House of Representatives on November 17, 2020 aimed at plugging the existing gaps & addressing certain safety standards in the current aircraft certification processes with the Senate's Commerce Committee voting unanimously a day later on the pursuit of a similar legislation on November 18, 2020.

This has been followed by another investigation report on the Federal Aviation Administration (FAA) released by the U.S. Senate Committee on Commerce, Science & Transportation on December 18, 2020. The report, based on a 20 month long investigation by the committee, has uncovered a series of serious lapses in aviation safety oversight and failed leadership on part of the FAA. U.S. Senator Roger Wicker, the Chairman of the Committee, slammed the FAA in his scathing statement "It is clear that the agency requires consistent oversight to ensure their work to protect the flying public is executed fully and correctly"[18]

"Three Things cannot be long hidden; the Sun, the Moon, and the Truth"-
Buddha

Chapter 12

THE 'SCORE CARD' AT THE END OF 'FIRST-HALF'S PLAY'

"You win battles by knowing the enemy's timing, and using a timing which the enemy does not expect" – Miyamoto Musashi, The Book of Five Rings

After numerous, intensely fought, highly dramatic & adrenaline pumping showdowns and rounds of bloody wars & bitter battles fought much in the 'eye for an eye and tooth for a tooth' manner across battlefields & war theatres over decades, with some of them won by Airbus while others going Boeing's way; it is still difficult term any one of them as the absolute victor in these end-less battles & eternal quest for absolute supremacy as there is no final battle or race lap which could be termed as the last battle or lap and no championship as the last & decisive one in nature. It is akin to Le Mans ad infinitum and an eternal testing of endurance, consistency & the ability to stay on top of one's game while ever adapting to the potential changes in the game's rules, technology, tracks & the competitors in the quest to outgrow & outperform one's own self while ever looking to outmaneuver & outclass the opponents simultaneously.

This is given that the organizations, unlike human beings, have this remarkable privilege of not having to ever leave the paddock by hanging up their boots to retire from the game ever and can always renew themselves and can always stage a comeback being more prepared, equipped & better armed for the next round of battles to come.

Therefore, a victory in such kind of a competitive game paradigm & structural layout could never be termed as absolute and can only be described as relative and transient at best while riding slowly on the ever so relentless wheels of time till the next championship takes over & becomes relevant while the fleeting glory of victory fades away finding its way into the record books.

Airbus and Boeing, both collectively have had an indelible mark, having shaped the evolution of commercial aviation over the past quarter of a century, with Boeing's development of the 707 commercial jet airliner in the 1950s followed by the development of world's first wide body aircraft; the 'queen of the skies', the 747, in a span of mere 16 months in the 1960s; being its trailblazing efforts in commercial aviation (besides the 4 Collier trophies Boeing has won for the 747, 757/767, 777 & the 787).

Airbus' list of contributions, in comparison, would surely be topped by the introduction of path breaking, fly by wire & flight envelope protection technologies, on the A320 in 1980s and the introduction of world's first pair of wide body twins, the A300/310 in the early 1970s.

Airbus and Boeing duopoly has seen the positions, titles & trophy change hands multiple times in the past across laps, races, circuits & championship titles but the thing that has remained constant throughout has been the sheer scale, nature & bitterness of the sworn arch rivalry that has been extremely well preserved and how it has played out over the decades despite having different players on the ever evolving, slippery &

treacherous tracks. Airbus, however, starting off as a rookie in the game positioned way back in the starting grid has put up a tremendous display of skill, determination, courage, grit & resilience clocking lap over lap looking to hunt down the reigning champion, especially in the narrow body league, while driving nonchalantly on slippery surfaces & bumpy tracks even under torrential conditions ever eager & always looking to pick up invaluable points for the championship title while waiting patiently under Boeing's rear wing on numerous occasions for the perfect opportunity to make the pass as the old master stumbled over some really tight & bumpy corners of the track.

And Airbus, since overtaking, hasn't put a wheel wrong so far on the track and has guided & navigated its way skillfully & carefully to a relative, transient championship victory by the narrowest margin for the time being in the narrow body league while Boeing has literally been forced into the pits for an impending overhaul following a major frontal crash..

The First Half's Play – Halftime Review

It will be fair to treat the current year, 2020, as the almost the end of the first half of the game to draw fair comparisons, evaluate performances, award points & to choose the victor for this first half by treating this as the ephemeral finish line & the referee's whistle marking the end of the first half's play as chosen by the force majeure. This is given that the COVID-19 has been the worst ever & unprecedented crisis in the past full century that has almost forced the entire industry into pits by having grounded a huge part of the global in-service commercial aircraft fleet across regions. The first half of this intense aviation power play has been marked by some great sporting & product strategy action through the decades.

1980s Decade: A Shared Decade

The great aviation power play started in the 1980s as prior to that it was an all Boeing aerobatics show with Airbus having just appeared in the early 1970s and undertaking the warm-up lap through the 70s decade with McDonnell Douglas' sporadic commercial cameos.

The 1980s witnessed the start of the race with Airbus launching the A320 in early 1980s looking to capitalize on airlines deregulation in the U.S. market in 1979 with Boeing also leading the way on the same path having launched its second generation of 737, Classic, of which the first variant, 737-300, entered service in 1984.

Boeing had also established its middle kingdom with the 757 & 767, both of which were launched in late 1970s and entered service in the first half of the 1980s decade followed by a botch-up & ultimately an aborted launch on the 7J7 during the second half.

Airbus launched the A330 & A340 in parallel in the second half of the 1980s as derivatives of the A300/310. The A320, took a technology leapfrog with fly by wire & flight envelope protection, entering service in 1988 with a disastrous take-off but had a perfect cruise phase subsequently with Boeing realigning compass & switching gears to go cruising in the wide body landscape leaving Airbus to create deep inroads & entrench itself perfectly in the U.S. market setting the stage for a lasting duopoly.

Overall, a shared decade with Airbus getting off the launch pad in the narrow body market successfully while Boeing got the 737 Classic and the 757 & 767 strategy right.

1990s Decade: A Boeing Decade in the Wide-Bodies and the Rise of Airbus as an Equal Rival propelled by the A320 Program

The 1990s witnessed the Boeing's 777 successfully getting airborne by the middle of the decade, taking on the A330/A340, while the 737NG, the most successful of the 737 generations; also took off to the skies. The twin-engine 777; launched as the first wide-body with early ETOPS backed by some serious statistical MacGyvering; turned out to be a huge success for Boeing while the A340 went off-course going with four engines in the ETOPS-180 era.

The second half of the decade saw Airbus going for another strategy blunder with the A380 quadjet while Boeing dabbled with exotic concepts, led by the larger stretches of the 747, as a counter plan while exploring the point to point model. The 1990s undoubtedly was a Boeing decade as far as wide bodies are concerned.

However, the 1990s was also the decade in which Airbus managed to take off to the skies with the A320 program backed by multiple, big ticket aircrafts deals with airlines while countering Boeing's market hegemony effectively and becoming an equal rival by the turn of the century.

2000s Decade: Another Split One

The Airbus-Boeing arch-rivalry almost reached its bitter peak in the 2000s with the decade turning out to be a real 'wide' one for aviation with a total of 4 clean sheet, wide body programs getting developed by the duo through the decade with the transatlantic rivals taking a contrasting approaches to product strategy.

Airbus took the conventional route with the A380 entering service in the second half of the decade while Boeing opened the point to point model with the 787 while also opening the second front against the A380 with the stretched 747-8 derivative.

Airbus reluctantly had to follow up with the A350 in the second half to take on the 787. Of the four programs launched by the duo, the A350 has been the only one to reach the breakeven zone so far while the 787 though has been deemed a market success with 1500+ orders but still a long way away from even break-even.

This was also a decade in which global airlines were in red in almost 7 of the 10 years of the decade (200-2009) marred by multiple, major crises led by 9/11, H1N1 and global financial crisis of 2008. At the end of the tumultuous decade; another split verdict for the duopoly…

2010s Decade: An Airbus Decade to the Core

The 2010s saw a stupendous comeback and truly belonged to Airbus with the company firing the first salvo with the launch of the re-engined A320neo in December 2010 while Boeing retaliated months later with the ill-conceived 737 MAX plan.

Airbus also invaded & captured Boeing's middle kingdom successfully with the stretched A321LR & A321XLR variants with Boeing at wits' end with scattered & fully engaged forces on other fronts and thus ended up with nothing to show up in return. The other product strategy masterstroke from Airbus came in form of the C-Series acquisition from Bombardier; a move which in fact could have been Boeing's truly wild card for the narrow body segment.

Another key angle to view has been the contrasting aircraft production & delivery rates for the duopoly for the second half of the decade (as the first half has already been covered earlier). For the 2015-2019 period, Airbus' aircraft delivery rates increased consistently from 635 for 2015 to 800 for 2018 & 863 for 2019 registering a 26% increase from 2015 to 2018 and 36% for 2015-2019 with the neos leading the charge. Boeing's assembly lines, on the contrary, had a breather in this phase (from 2015 onwards) with aircraft deliveries growing by just 6% over 2014-2018[9] following the softening of order intake on the wide body front; with protectionism & trade wars coming into play; and a subsequent rate adjustment by Boeing, especially on the 787 & 777 programs, while the 737 deliveries continued to increase through this period. The 777 deliveries almost halved from 99 aircrafts for 2016 down to 48 aircrafts by 2018.

The other action coming in the decade, which more or less belonged to the narrow bodies, has been the launch of 777X, roll out of additional 737 MAX & 787 variants by Boeing and the NMA pipe dream. Overall, well played Airbus!

Narrow Body War Front: The Inflection Point

The most remarkable moment & the inflection point in this old & bitter sworn arch-rivalry in the narrow body market came when the original challenger finally overtook and edged ahead of an exasperated old master on one of the familiar series of corners of the tricky narrow body Nürburgring…

That happened in October 2019 when the cumulative order tally of the Airbus A320 family overtook Boeing's 737, grounded & crippled since the MAX disaster. Airbus cumulative order book for A320 family stood at 15,193[7] aircrafts as of the end of October 2019 while Boeing's 737's cumulative tally was 15,136[7] units with Boeing's continued lead in overall

---

[9] Factoring 2019 numbers into analysis won't make sense as the 737 MAX was grounded in March 2019 and no deliveries were made for the rest of the year on the program.

narrow body aircraft deliveries over its arch-rival being the sole saving grace for the old master which has so far found itself descending further down the self-created rabbit hole. Airbus' technology & customer focused overall strategy orientation in the narrow body aircraft market, thus, has comprehensively prevailed over Boeing's competition oriented reactive strategy focus at the end of first half's play.

As of the end of year 2019, Airbus has been in the commanding position in the narrow body aircraft segment having won the sort of 'triple-crown of the aviation's narrow body league' with the A220 in lead at the bottom of the pack with a clear head start, the A320neo firmly in command in the middle and the A321neo-LR & XLR dominating the skies while giving the top cover with Airbus having almost wrested the control & crown of the narrow body market from Boeing.

The segment specific market shares and the order book position in the narrow body market at the end of 2019 so far are skewed towards Airbus giving much discomfiture to Boeing's well known hubris & its usual disdain for the European airplane-maker. Riding a strong, commercial aviation demand growth wave, Airbus' total order intake for the A320 aircraft family (since inception of Airbus) stood at 15,315 orders[1] (split 53:47 amongst the ceo & neo variants) as of December 31, 2019.

Compare that to the Boeing's 14,969 cumulative firm orders[2] for the 737 program (including the numbers for the 737 Original & Classic series) as of December 31, 2019 with hardly any decisive difference between the two sides to choose from with the only differentiating fact being that the A320 took off almost two decades after the 737.

For the A320 family order book, around 7,151 orders belong to the A320neo family, of which 3255 belong to the A321neo variants alone (almost 45%), which is the very zone that has made the key difference over the past decade and has created the edge for Airbus with the A321neo variants having captured the middle of the market in this decades long aerial marathon.

The order intake has been taken as a preferred surrogate measure of assessment as it reflects the market's responsiveness & verdict on the OEM's product strategy moves just as the stock price's directional movement is taken as the market's barometer of the assessment of a company's moves & performance.

<u>Illustration: 1</u>

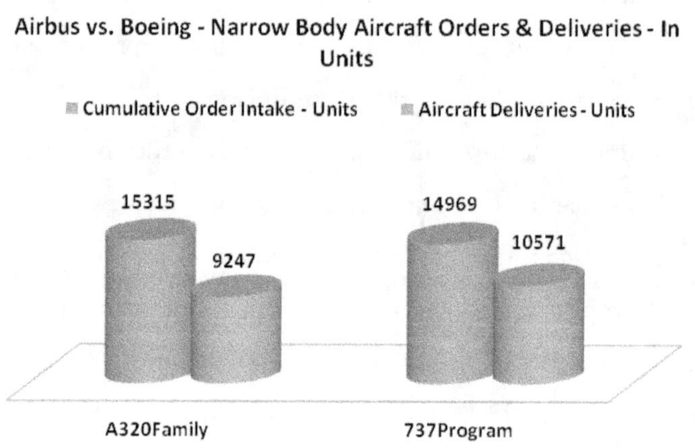

Thus, cumulative deliveries for the A320 aircraft family (since its EIS in 1989), stand at 9,247 units (including the voracious neos) while the order backlog stands at 6,068 aircrafts[1] as of Dec 31, 2019, thereby, positioning the European plane maker firmly in command on a very strong growth trajectory for the 2020s decade.

The A320 deliveries have again been very close to the Boeing 737 program's 10,571 cumulative aircraft deliveries (since achieving EIS 20 years earlier than the A320 in 1967) and 4,398 aircraft strong firm order backlog as of the end of 2019[2] with the MAX numbers impacted by its ignominious grounding since March 2019.

For Boeing, the 737NG has been the most successful 737 generation with 7,110 aircraft orders and 7057 deliveries followed by MAX with 4,741 aircraft orders[3]. Boeing's order intake count for 2019 was -89 with 205 order cancellations for MAX worth $10.6 billion, while Airbus booked 717 new narrow body orders worth almost $40 billion, as per Airbus & Boeing data[5,6]. In terms of deliveries, Airbus delivered 863 aircrafts while Boeing shipped out only 380 wide body airplanes amid ongoing MAX grounding.

Illustration: 2

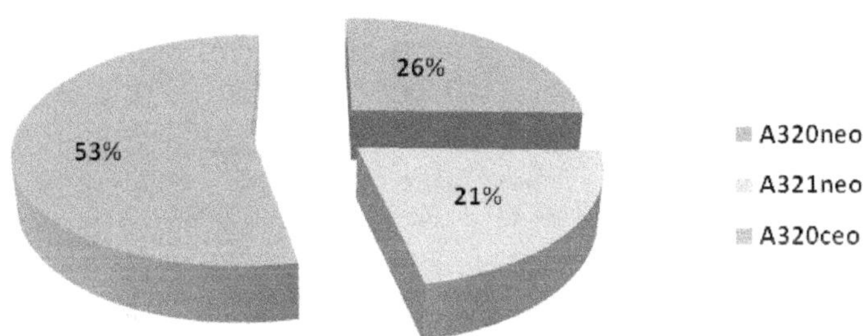

Airbus - Order Intake Split by Aircraft Families - Since Airbus Inception to Dec 31, 2019 - Total Orders - 15, 315 A320 Family Aircrafts

Illustration: 3

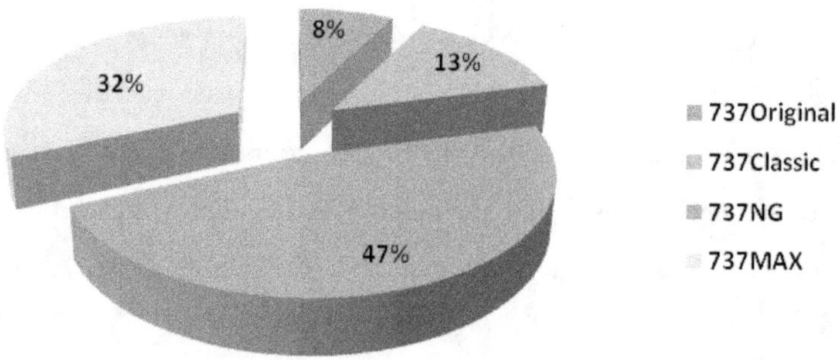

Illustration: 4

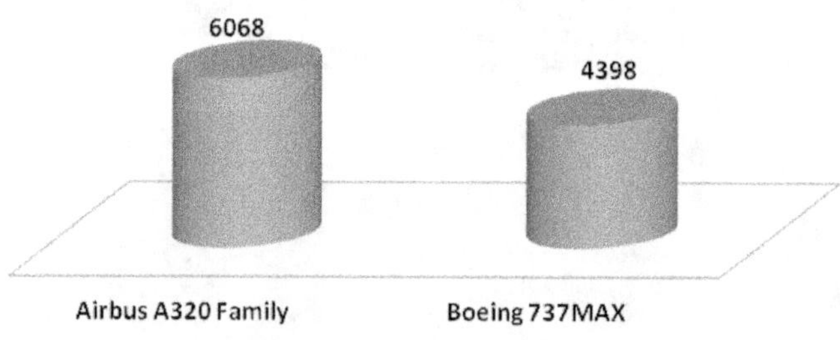

The narrow bodies, led by the A320 aircraft family, overall, thus, have collectively accounted for almost 80% of Airbus' cumulative order intake of 20,108[1] aircrafts (including the A220) as of December 31, 2019; with the

A320 family accounting for 76%; while the wide body programs (A300/310, A330, A340, A350XWB & A380) collectively account for just 20% of total order intake (with just under 5,000 wide body aircraft orders) since the company's creation!

Boeing's side of the picture has been relatively much more balanced with a total of 25,000+[2] orders, as of December 31, 2019, of which 60% belonged to narrow bodies (including 727 trijet) while the wide bodies accounted for a 40% share of the order intake, led by the 777, 747 & 787 programs.

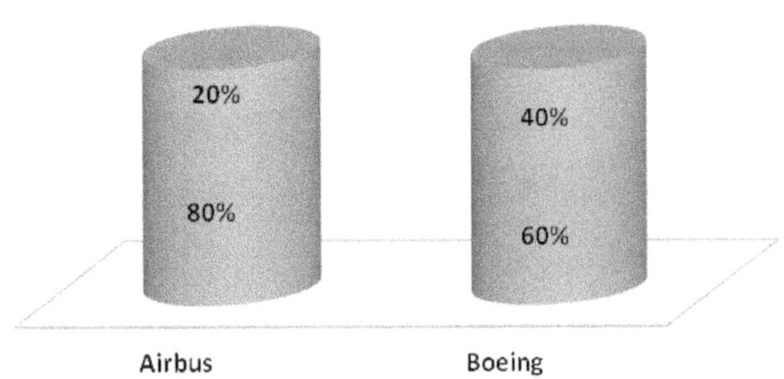

The other interesting aspect is that the A320ceo family grew to the 8,000+ order intake level in almost 3 decades (1989-2020) while the A320neo family has taken less than even a single decade (since A320neo launch in December 2010) to be almost there with 7,000+ orders (exactly 7,151 with A321neo variants accounting for 45%) already captured by December 31, 2019[4]. Contrast this with the Airbus' original conservative estimates at launch in December 2010 with Airbus having projected a paltry 4,000 orders for the A320neo family over 15 years[4]!

Airbus' strategy has been simple and is much akin to the one usually pursued & seen in the Ashes test cricket championship fought between bitter rivals, the English and the Aussies. Like the ashes, herein, Airbus, the dominant side in narrow bodies, would invariably bat first with a proactive strategy focus & a strong line-up and would put up a good total on the board and would put pressure on Boeing with a good line & length bowling attack along with superior tactical execution.

This would pivot the game towards two potential outcomes, in terms of probabilities and irrespective of the surfaces & ground conditions; whether swinging, bouncy or spinning; either towards a Draw if Boeing bats well or an Airbus Win if Boeing buckles down to pressure tactics.

Airbus' loss would be possible if it fairs poorly while batting in the second innings which it has usually avoided, at least in the narrow body segment, with a calm & calculating demeanor while already having avoided a lead build up by Boeing in the first innings followed by the pursuit of a well laid out strategy course & detailed scenario planning. The narratives in the wide body segment though have been entirely different with Boeing usually starting in the lead and Airbus more prone to errors.

The A320 aircraft family, despite having started a good two decades later than the 737, has been able to surpass the reigning champion in many qualitative aspects of the game given that in terms of numbers it has always been a neck in neck situation and almost a photo finish between the two sides and it actually is too close to call.

However, evaluating & judging it from the qualitative aspects; led by, overall strategic responsiveness, technology innovations & advancements and overall contribution to aviation's evolution in the narrow body operating context; Airbus could deservingly claim its right on the narrow body league title at the end of first half by the slimmest margin along with the podium in a virtual nail biting cliffhanger.

In the light of the wars & dogfights fought bitterly between the two aviation giants over the past 5 decades, covered amply in the previous chapters, it will be fair to award the narrow body trophy & the champagne to the Airbus team for this first half's play at half time with the confetti flying in the background. As discussed earlier, it will be fair to treat the current year, 2020, as the almost end of the first half of the game.

The Airbus victory, predicated on the A320 family, has been scripted by a blend of visionary approach to strategic decision-making focused on the long term, proactive strategic focus & high strategic responsiveness (marked by nimbleness & agility to changes in environmental factors), significant hard work on tough tracks & back in the pits with brilliant tactical execution and the eagerness to ever stay on the cutting edge of technology while simultaneously maintaining a firm grounding deep rooted in hardcore reality that has enabled the company's transformation from starting out as a Rookie way back in the early 1970s (wide body segment) and late 1980s (narrow body segment) to a deserving Champion almost 3 decades later in 2020.

The next decade, i.e. the 2020s, most likely is going to go Airbus way in a post COVID-19 world with a successful A320 line-up likely to go all out again to outgun an embattled & caged 737 MAX under close scrutiny and still far away from being able to trace its way back up to skies.

The Neo vs. Max battle is likely to be stacked more in favor of the way better positioned Neo going forward with the Max badly required to re-establish its credibility & re-validate flying credentials in a virtually uphill battle.

The addition of the A220 to Airbus portfolio broadens portfolio and positions Airbus in a commanding position at the lower end of the narrow body segment while Boeing has messed up on the Embraer deal. The A220 also provides Airbus the option to jet out its next scratch-up narrow body in a jiffy and almost waiting in the offing to outfox & outclass any potential rivals from the Boeing stable.

The long-legged A321XLR already is dominating the middle of the market league led by its outstanding operating economics with an 80%+ market share while the outdated, in-service 757s languishing on their slothful journey towards obsolescence & perdition.

Boeing, on the other side of the pits, is literally on the ropes in a dark corner of the ring and has a big task ahead for itself with a multi-front, existential war to be waged & won on the toughest of terrains under full regulatory & public spotlight as well as scrutiny against a sworn arch-rival under volcanic operating conditions marked by a historic demand meltdown.

With the wide body aircraft segment, Boeing's traditional forte, in absolute dire straits and a grounded & chained 737 MAX (for over a year now) unable to make things any better in the narrow body segment; Boeing's options are limited while facing & addressing a crescendo of charges & demands from airlines & public alike for compensation making the situation descend slightly towards a potential financial distress in the making.

Getting the runway cleared for the 737 MAX's take off roll will no doubt be on top of the list for Boeing followed by finding a well thought solution to the A321XLR conundrum through a sorted & well charted out product strategy while carefully avoiding falling into the typical trap of knee jerk reactions completely. The resumption of flight tests of the 737 MAX towards re-certification in early July 2020 under berated FAA hawks has been the sole silver lining for Boeing so far in 2020.

A relatively much better business portfolio diversification, led by a stable defense & security business, does give Boeing a competitive advantage over Airbus along with a stronger strategic & market positioning in the wide body segment, especially, the relatively superior positioning in the cargo aircraft market.

Airbus, however, has a relatively better & stronger balance sheet and cash generation profile at the moment along with much lesser investment requirements with a more or less settled product portfolio in the narrow body segment while Boeing clearly needs to plug the gaping hole in its portfolio to effectively address the middle of the market segment to counter A321XLR's hegemony with a 757 replacement while also contemplating an eventual succession plan for the 737 MAX.

In terms of culture, leadership and strategy; the three key determinants of long-term organizational success; Boeing had its original engineering culture diluted & eroded post the acquisition of McDonnell Douglas (as covered in part 1), which was in turn supplanted by a profit focused culture. Boeing also had serious leadership issues for the 1996-2005 decade with two of its chairman & CEOs and a CFO ousted over scandals & ethical aspects. Under McNerney's reign spanning 2005-2015, Boeing ravenously hunted for profits in the overdrive mode supplemented by a low-investment, low-risk derivative product strategy which was preferred by his successor Dennis Muilenburg as well.

Airbus, on the contrary, has persisted with its original customer-focused culture; instilled as a core value by Roger Beteille, one of the company's founding fathers, in 1969; for almost half a century now. The Airbus deck, however, had been marred by factionalism emanating from its split leadership model based on nationality which undermined its efficiency, nimbleness & competitiveness till almost the end of the first decade of the twenty first century. In terms of strategy orientation, Airbus' has mostly been oriented towards customers & technology, especially in the narrow body segment, which led to A320's successful market disruption with the FWB technology in the 1980s and the creation of A321LR & XLR variants.

Boeing, on the contrary, has largely been on the reactive with its product strategy mostly oriented towards competition in the narrow body segment as highlighted by the creation of successive generations of the 737 over the past decades driven largely by competitive pressures.

The second half of the game, likely to be played out over the next few decades, thus, is likely to be an interesting contest if Boeing learns from its mistakes & errors and bounces back as a revamped & completely overhauled unit in the narrow body format as well, giving its highly overdue share to it and Airbus continues to maintain its lead in this high octane perennial contest for aerial supremacy.

The most important change which we could probably see going forward into the future is that it is likely to become a 3-team league once again with COMAC clearly being the next rookie standing in line and China's relatively much better maneuvering of its way out of the COVID-19 pandemic.

China's determined push for the development of a full-fledged domestic aerospace industry for the world's largest aviation market backed up well by deep state pockets and with a no direct confrontation strategic posture means that dispensing off with COMAC is not going to be that easy for the duopoly going forward as it has been with an impulsive Bombardier.

COMAC, thus, is likely to stick around with low market shares for long trying to take the game down to the wire before switching to the wet mode with afterburners and going for the full throttle charge. The Chinese aviation market's scorching growth rate; with the country accounting for a mere 2% of global aircraft orders in 2001 to 23% by 2018 and almost a third prior to the COVID-19 outbreak; is a clear testimony of China's overall capabilities and what's still to come on the road ahead.

However, being human beings, many of us may not still be there in this physical existence of ours to be able to see that grand 'Super Bowl' finale of the season…The onus of writing that exciting sequel to this one covering the second half's play, which will most likely be doable by the mid-century or may be a decade or two later from then, may fall on the next generation…

"If you do not control the enemy, the enemy will control you" – Miyamoto Musashi,
The Book of Five Rings

Chapter 13

THE 'LAST SORTIES' BEFORE THE 'SITZKRIEG'!

"I never see what has been done; I only see what remains to be done" – Buddha

One of the longest aviation growth cycles of commercial aviation; having taken off in 2004; came to a highly unexpected & abrupt grinding, screeching halt dramatically at the turn of the decade in March 2020 post the COVID-19 pandemic outbreak. Almost the entire global, in-service commercial aircraft fleet was unprecedentedly forced towards undertaking simultaneous touchdowns & rollouts in March 2020 leading to aircrafts jostling for parking space across airports & other makeshift storage sites worldwide. The COVID-19 tsunami and its high magnitude aftermath seem to have eclipsed the other raging issues & key trends being faced by commercial aviation, which include:-

1. Sustainability Challenges & the Journey towards Flying 'Carbon Free'

The growing concern about our carbon footprint and its impact on environment has been one of the top concerns facing the world today. Air transportation has been a key part of the complex global warming conundrum and with the scathing activist campaign against air travel gaining significant momentum & almost on the verge of becoming a social movement; spearheaded by the next generation being led effectively from the front by Greta Thunberg as the flag-bearer, whose September 2019 global climate strike waged across 163 countries drew in almost 4 million[1] people worldwide marching collectively under the guidon of 'Greta Force' and its 'Flygskam' motto. There has been a tremendous & unprecedented pressure on the industry, regulators & governments across the globe to undertake serious, actionable measures to address the situation before it reaches escape velocity and the humanity is beyond the point of no return.

The aviation industry's contribution to total global greenhouse gas emissions has been pegged at around 2% or 860 million metric tons of CO_2, which is projected to grow further by 300% to 700% by 2050 as against the levels prevailing in 2005, as per International Civil Aviation Organization (ICAO).[2]

The International Air Transport Association (IATA) intended to achieve carbon neutral growth under its 2013 resolution on Aviation Carbon-Neutral Growth (CNG) 2020 Strategy with plans for a further 50% reduction in net emissions by 2050 as against the 2005 levels. About 40% of the expected reduction was likely to have been achieved through the use of alternate fuel based propulsion systems with the rest attributable to technological developments followed by pursuit of improvements in technology, operations and infrastructure development with deployment of single, unified global Market based Measures (MBM).

The aviation industry needs to invest significantly towards development of new technologies, especially, in the development of a range of sustainable alternate fuel based technologies to achieve the laid out objectives amid a difficult global macroeconomic environment, especially, in the post COVID-19 scenario.

Stopgap Mechanisms Crafted by the Industry – Carbon Offsets

The industry eco-system has been working towards boosting production output of sustainable aviation fuels (SAF) as one of the strategies to meet the emission targets over long term with the industry likely to reach the tipping point on this by 2025.

Another interim, mid-term mechanism being worked out by the industry is Carbon Offsetting and Reduction Scheme for International Aviation (CORSIA), which is a market based mechanism which will enable airlines to offset carbon emissions produced by international air travel, as per ATAG, a cross industry association comprising industry bodies like IATA, Civil Air navigation Services Organization and Airports Council International in addition to industry OEMs & engine manufacturers.

CORSIA is scheduled to be implemented phase-wise from 2021 onwards and is likely to provide around $40 billion for climate projects which would be able to offset around 2.5 billion tons of CO2 over a 15 year horizon[3]. Around 80 states have so far signed up for CORSIA which will be able to cover about 80% of the projected growth in international aviation emissions over the next 15 years with further support over near term critical to the mechanism gaining critical mass.

IATA launched the world's first Aviation Carbon Exchange (ACE), a centralized marketplace for trading of CORSIA eligible emission units backed by IATA Settlement Systems and Clearing House, in collaboration with Xpansiv CBL Holding Group (XCHG). ACE was unveiled in late

November 2020 with JetBlue Airways becoming the first airline to make a transaction on the ACE platform while also creating history in the process.

However, the effectiveness of offset mechanisms in delivering the claimed reductions has traditionally been questionable. "The research shows that three-quarters of the offsets don't deliver the reductions they claim to deliver[2]," said Anja Kollmuss, a Zurich based policy analyst studying emissions trading. The CORSIA's effectiveness in packing a punch on emissions, thus, could also be somewhat along the traditional lines!

2. 'Electrify Thy Flight': Industry's Experimentation with Electric & Hybrid Propulsion Systems for Aviation Applications

Electrification has been the second pathway being pursued actively in the aviation's nascent & long journey towards sustainability, which has more or less been the story for other members of the transportation industry, led by Automobiles and Trucking, which have made rapid & tremendous progress towards it. Electrification had been preceded by the development of biofuels, which beginning as sporadic efforts across continents, somehow have been facing major challenges on the scalability part of the equation.

On the electrification front, the industry has been taking small, nano steps with the power density limitation of Lithium Ion batteries, as against conventional jet fuel, posing serious limitations for the scope of aviation specific applications it can support. Projections indicate that at the current rate & pace of battery technology's evolution, it will only be by the middle of the current century that batteries would have become capable enough to make a serious propulsive difference to the aviation's journey towards mainstream & commercial sustainability beyond the current brigade of concepts & niches.

Amongst key & notable industry efforts towards electrification; the Airbus, Rolls-Royce and Siemens Teaming Up on the development of a Hybrid-Electric Propulsion System for commercial aviation; dubbed Project E-Fan X[4], has been the most serious effort to date. The three industry behemoths have been collaborating with initial plans looking to get to fly a demonstrator aircraft through 2020. Under the project, Airbus had been responsible for the control architecture of the hybrid-electric propulsion system as well as batteries and integration with flight controls, Rolls Royce has provided turboshaft engine and the 2MW generator while Siemens was to produce the 2MW electric motor for the propulsion system.

The Project E-Fan X achieved a key milestone in early 2020 with the successful completion of the wind tunnel testing of the BAe 146 R100 hybrid-electric demonstrator aircraft's scale model by Airbus. The demonstrator aircraft was to have one of its four jet engines replaced by a 2MW motor.

The E-Fan X was scheduled originally to undertake its maiden flight in 2021 prior to which flight tests for characterization & installation of the 2MW motor on the test aircraft were to be completed towards late 2020[4]. Airbus and Rolls Royce have been in the process of testing the individual components of the hybrid-electric propulsion system. Airbus also recently inaugurated its latest dedicated test facility, based at Munich, in October 2019 for testing full and hybrid-electric aircraft technology. The facility has been named as the E-Aircraft Systems House and was to conduct comprehensive testing of the individual components of the hybrid-electric powertrain to be used on the E-Fan X demonstrator.

However, the project, as with many other future focused industry projects, got swept away in the economic aftermath of the COVID-19 tsunami which has decimated commercial aviation for years to come. Amid the pandemic induced market carnage and the dawning of the realization that battery electric is going to be a long term play hinged on technology evolution, the industry has looked towards hydrogen based propulsion in the interim; mirroring & much on the lines adopted by trucking as well with the potential added advantage for a hydrogen powered aviation sector being the centralized infrastructural requirements which makes for a great structural fit with hydrogen's overall infrastructural set-up as well.

The French government's industry bailout package unveiled post COVID-19 includes the ambitious plan to develop the world's first hydrogen powered commercial airliner as a replacement for the A320 by 2035 now, instead of 2050 earlier, which if incarnated by then would enable Airbus and the European industry value chain to raise the bar by a huge margin and going to provide a definite head start in the Aviation 2.0, post a major reboot under the hood, in a world which hopefully will be more sustainable than now!

3. The 'Supersonics' Strike Back

The Supersonics are back in vogue since disappearing from the skies in 2003 with the retirement of Concorde. However, this time round they are going to be making their special appearance first in the business aviation league rather than commercial with the charge spearheaded by Aerion Supersonic, who has been planning & developing a family of supersonic aircrafts. Aerion's flagship aircraft, AS2, will fly at high sub-sonic speeds over land and will switch to the supersonic mode over the transoceanic leg of its flight.

The Aerion's supersonic aircraft project[5], scheduled to enter service in 2024, incorporates a low-sweep, thin wing design aimed at delivering high fuel efficiency while assuming no changes in regulatory guidelines and envisages a combination of optimum sub-sonic speeds over land and supersonic speeds over oceans for aircraft's operations to reduce the journey time by almost half over a typical transatlantic flight. The city to city connecting flight between New-York to Paris is likely to take 4 hour & 15 minutes with the AS2 as against 7 hours & 30 minutes at present.

GE Aviation has been tasked with MacGyvering the propulsion system for the AS2 based on its existing CFM56 based engine core[5]. Boeing has also joined the supersonic league indirectly with the announcement in February 2019 that Boeing is partnering with Aerion and has made significant investment (amounting to several hundred million dollars for a 40% stake) in the company as Boeing places its bets on supersonic air travel. Boeing will be providing Aerion with engineering, manufacturing and flight-resources, in addition, to strategic vertical content for the AS2 program. Boeing replaces Lockheed Martin Corporation as program partner on the AS2.

However, current regulatory guidelines prohibit supersonic flights over land, by the FAA in the U.S. & by the International Civil Aviation Organization (ICAO) internationally, on account of the sonic boom issues. The regulatory approvals & changes are likely to be implemented & come into effect only once requisite technologies for mitigating sonic boom effects to acceptable levels are developed, tested & matured, as has happened with all technological revolutions across industries & sectors. The industry has already been at work over this under the guiding beacon of NASA.

Aerion has been working diligently on supersonic technologies for last 16 years with the company having filed multiple U.S. patents pertaining to supersonic technologies, including, multiple patents for laminar flow aerodynamics which Aerion terms as the key to the development of AS2's wing. Aerion has also partnered with NASA over recent years to develop, test & refine many of its laminar flow concepts and technologies.

The key difference in the AS2 lies in the manner in which the sonic boom is handled. As per Aerion; unlike traditional aircrafts, where the sonic boom is mitigated with aerodynamics; in AS2, Aerion plans to use boom mapping software to mitigate the boom which will enable overland cruise speeds of up to Mach 1.2 by ensuring that the boom generated does not reach the ground.

The AS2 program is currently in the preliminary design phase which is scheduled to be completed by late 2020 with the maiden flight slated for 2024. Aerion expects the supersonic business jet demand to be for 300 aircrafts over 10 years worth almost $40 billion on the base of overall program development costs of around $4 billion[15]. Others who have also been on the supersonic pathway at some or the other stage of development, include, Gulfstream, which has also filed some patents pertaining to supersonic flight technologies.

However, whether we will see a commercial supersonic airliner riding the transatlantic jet streams again into the in future is the question to be asked. Given the airliners financial stinginess & fixation with operating economics, growing sustainability challenges & ongoing activist campaigns against flying and the economic fallout of COVID-19; that possibility seems obscure in the typical format. But there is a crusader & market challenger trying to defy these odds and is building the commercial supersonic aircraft of the future trying to disrupt the market by going niche & leveraging exotic technologies.

Boom Supersonic has been developing a futuristic commercial supersonic airliner trijet, christened Overture[6], capable of flying at top speeds of up to Mach 2.2, courtesy GE Aviation & its tremendous engineering prowess, which is providing its medium bypass ratio J85-15 engines to the program.

Boom has announced that it will be unveiling its scale demonstrator in October 2020 (despite COVID-19) followed by flight tests likely in 2021[6]. To counter sustainability challenges, Boom plans the Overture to be solely powered by 100% sustainable aviation fuel (SAF) with the $200 million supersonic airliner[6] likely to enter service around 2030 to take us on the long awaited commercial flight beyond the sound barrier!

4. Boeing's Invasion of Supplier's Turf under its Vertical Integration Strategy

Boeing has been encroaching on the supplier's territory since 2014, taking subtle jabs at suppliers, looking to maximize on the market's upswing phase, as part of its vertical integration strategy, by in-housing multiple aspects of aircraft development with some mirroring of the strategy also done by Airbus.

Boeing continues to broaden its vertical integration strategy marked by significant in-sourcing of work areas traditionally performed by T1 suppliers. The swing in industry trend towards resurgence of vertical integration now follows the much vaunted decade of the 1990s, which witnessed the onset of outsourcing and creation of Global Tier-1 supply chains.

Key drivers of this in-sourcing strategy have included:-

- Huge order backlog for new aircrafts across Boeing & Airbus valued at over $1 trillion. The OEMs had been looking to capitalize on & leverage this long term visibility of the order book position to make investments towards production of aircraft equipment and generate returns on that investment

- Gaining control of areas with strategic interests to ensure no disruptions to the production schedules originating from suppliers and as a protective measure to counter the growing clout and bargaining power of suppliers given the ongoing wave of consolidation in the industry marked by some mega mergers over the recent years

- Leveraging the clear visibility of its order book position of commercial aircrafts through the next decade indicating significant revenues growth potential from expanding production base from simply final assembly to producing sub-systems as well

- Achieving cost reductions on existing aircraft programs to boost profitability while bolstering overall efficiency, shortening development horizon and meeting turn around & delivery timelines

- Bolster overall profitability given the higher profitability of components & sub-systems as compared to aircraft assembly. Profit margins for the aircraft & engine OEMs have been under high single digit percentages over the recent years as compared to double digits for the Tier 1 & Tier 2 suppliers. The decisions by the aircraft OEMs going for vertical integration therefore seem obvious.

- The OEMs looking at the steady, long term cash flow streams which typically originate from servicing & MRO activity over long term which is significant & is capable of offsetting the typical cycles involved in commercial aircraft sales. Boeing had publicly stated that it intended to grow its services business to $50 billion by 2022.

- To somewhat offset what Boeing believes as lack of value creation from the UTC-Rockwell Collins merger earlier in 2017 and to counter its suppliers becoming too big and gaining a negotiating clout & heft in the industry. Boeing has clearly been targeting UTC as far its in-sourcing initiatives are concerned. Earlier in 2012, when UTC acquired Goodrich, Boeing had in-sourced its nacelle production and switched suppliers of the 777 landing gear. The supplier consolidation rally in commercial aviation has been led by United Technologies in the U.S. and by Safran across the Atlantic in Europe.

The move however entails significant risks for the suppliers which are already feeling the squeeze and have not taken it well. It may also further accelerate industry consolidation. The move, however, has deepened Boeing's presence in the aircraft equipment market under its BGS (Boeing Global Services) brand that has been giving jitters to the suppliers in the global commercial aircraft market with the aviation giant straying into their territory.

Key milestones on Boeing's in-sourcing journey have included:-

Boeing's JV with Safran to manufacture APUs for Commercial Aircrafts – June 2018

Safran Group & Boeing made a key announcement for establishment of a JV for the production of Auxiliary Power Units (APUs) for the commercial aircraft market. As per the agreement, announced in June 2018, the two companies were to design, produce and service APUs for commercial aircrafts under a newly created joint company to be owned 50:50 by Boeing and Safran respectively.

The agreement, signed towards the end of 2018, significantly threatened Honeywell and United Technologies as they have been key suppliers of commercial aircraft APUs for decades now.

Honeywell is the sole supplier of APUs for the Boeing 737 MAX program apart from C-Series, Comac C919 & Irkut MC-21 programs. Pratt & Whitney, on the other hand, is the sole of supplier of APUs for the Boeing 787 program. APUs produced by both Honeywell & Pratt & Whitney, are offered as an option on the Airbus A320neo program as well.

In the wide-body aircraft market, Honeywell supplies APUs to the A350 & 777 programs while Pratt & Whitney supplies to the A380, 787 & 747-8 programs. Thus, the move is likely to impact Pratt & Whitney the hardest given the limited production rates of the A380 & 747-8 programs over the recent years and the end of the road ahead for A380.

Boeing's Acquisition of Aviation Spare Parts Distributor KLX Inc. – May 2018

Boeing made its biggest acquisition so far, as part of its strategy to create a $50 billion services business over the next decade, with the decision to acquire aviation spare parts distributor, KLX Inc. for $3.25 billion in addition to undertaking $1 billion of net debt valuing the entire deal

cumulatively at $4.25 billion. KLX has become part of Boeing Global
Services and has been integrated into Boeing's spare parts subsidiary, Aviall,
which was acquired by Boeing earlier in 2016. As per Boeing, the deal is
likely to generate annual cost savings to the tune of $70 million by 2021.

**Boeing's JV with Adient to Produce Seats for Commercial Aircrafts –
January 2018**

Boeing announced its foray into the commercial aircraft seating market
through a JV with Adient, the auto parts business spun out of Johnson
Controls in 2016. The announcement was made by Boeing in January 2016
and cited the move as being in response to the delays in seat production &
capacity constraints across current suppliers, namely, Rockwell Collins and
Zodiac Aerospace which have subsequently been acquired by UTC and
Safran respectively.

**Establishment of Boeing Avionics to develop and produce Avionics
and Electronic Systems – July 2017**

Boeing made the announcement to set up an in-house avionics and
electronics production unit in late July 2017 following years of outsourcing
the procurement of the same.

The move came against the backdrop of the mega industry merger with
Rockwell Collins lapped up by UTC. The decision has been a part of the
strategy to build targeted vertical capability. The new Avionics business
units will develop & produce systems for all Boeing aircraft programs and
not just commercial airplanes, which are likely to enter service over the next
decade. The business unit focuses on development & production of:-

- Navigation Systems

- Flight Control Systems

- Information Systems

- Other Core Avionics

Airbus, too, has been pursuing a similar strategy, however, with a much lesser aggressive intensity than Boeing. Airbus had announced for the first time in late 2017 that it plans to in-house some work pertaining to design of some engine nacelles which have been supplied so far to Airbus by United Technologies Corporation for the A320neo aircraft program for both the engine options offered by Pratt & Whitney & CFM International respectively.

Airbus' official stance & rationale regarding the decision has been touted as enhancing competitiveness with scope for improved efficiency and performance. However, there is a strong incentive behind the same for increasing profitability and to regain strategic control post the UTC-Rockwell Collins mega merger.

Airbus also had outlined plans to further produce thrust reversers currently produced by Pratt & Whitney and for integration of engine housing & the pylon. Airbus inaugurated the new engine nacelle integration plant for the Airbus A320neo based in Hamburg in October 2018 which had been set-up in collaboration with Safran, which was scheduled to ramp up production of the engine nacelles for the CFM International's LEAP-1A engine program by 2020.

However, Airbus has been quick to review vertical integration's overall scope and rescind on the nacelle part, moving out of it recently and banking on Collins Aerospace instead for that.

The COVID-19's impact on commercial aviation has been devastating with a recovery to the pre-crisis activity levels likely only by 2023 at the earliest. In the sudden & unprecedented turn of events, the investments made by Boeing towards in-housing have now become fixed costs & a

strain on the bottom line and will now need further extended time horizon
to recoup along with a rapid alteration of the now seemingly ludicrous $50
billion services business plan, further compounding the already long series
of problems facing Boeing post MAX fiasco.

5. The 'Bermuda Triangle' of the Narrow Body Ocean and the 'Disappeared Aircrafts'

The narrow body aircraft market does have a Bermuda triangle wherein
a number of airplanes have simply disappeared over the decades. And that
is the lowest end of the narrow body segment.

Historically in commercial aviation, the long stretching of the fuselages
to create larger aircrafts has witnessed much more success than the
opposite. The short stretching approach to create shorter aircraft variants of
the existing narrow bodies has failed miserably across both Airbus as well as
Boeing.

Airbus A318, the smallest variant & member of the A320 aircraft
family, was created by Airbus by short stretching the A319 with the aircraft
produced by Hamburg line entering service in July 2003. Capable of seating
107-132 passengers with an operating range of 3,100nmi (as per Airbus),
the A318 was also certified by EASA for steep approach operations making
it one of the largest commercial aircrafts to be certified for that, thereby,
opening additional application avenues for airlines in form of particular
airports, like London City.

The A318 program had its genesis in the joint Franco-Sino clean sheet,
proposed 100 seat aircraft program explored & discussed in the 1990s
which was ultimately grounded. Airbus, sniffing a market opportunity, thus,

went for it, shrank the A319 and positioned the A318 at the lowest end of the narrow body aircraft spectrum.

The program, however, never really took off the way Airbus had expected with only 80 A318s delivered by Airbus over the 2003-2013 decade with a significant number of them going to Air France. The A318 program encountered strong headwinds in form of a multitude of problems & challenges (some of them even structural), starting with Pratt & Whitney's new PW6000 engines coming short with issues pertaining to originally claimed fuel efficiency levels being reported and by the time CFMI came up with its CFM56-5; dovetailed by P&W's market revenant with the revised engine design; the A318 was already doomed with airlines having already switched to larger aircrafts.

Further, the regulatory guidelines across both sides of the Atlantic put the A318 in the same class as the larger, mainstream narrow body aircrafts, thereby, subjecting the smaller A318 to same bevy of airport charges as its larger counterparts impacting the program's market prospects directly as the airlines would have been unwilling to pay same charges while operating a much smaller aircraft necessitating higher operating frequencies. This was a structural problem without any headway.

One of the handful & very limited successful applications of the A318 has been by British Airways which has deployed the aircraft on the transatlantic London City-New York route with a refueling stopover built in between. The A318, thus, has been amongst the few misses Airbus has had in its otherwise largely successful A320 family, however, it still managed to perform slightly better than its direct rival from Boeing's stable, the 737-600.

The 737-600 was launched by Boeing as the smallest member of its

737NG series in the 1990s, as a replacement for the older generation 737-500 (which in turn replaced the 737-200) with SAS as the launch customer. The -600 entered service almost half a decade before the A318 in 1998 with a seating capacity of 108-123 passengers along with an operating range of 3,235nmi and since then has fared miserably in the market (facing similar issues as the A318 would face later) with a mere 69 737-600s[7] produced & delivered (with around 30 of the -600 aircrafts going to SAS alone followed by Canada's WestJet) by Boeing to airlines over the 1998-2006 period. Its predecessor, the older second generation 737-500 Classic also had the lowest production tally of the entire 737 2nd generation Classic series with 389 aircrafts[7] produced & delivered of a total close to 2000 for the entire 737 Classic series[7]. The aboriginal 737-100, the smallest member of the original 737 series, too had a similar fate with only 30[7] 737-100 aircrafts produced & delivered by Boeing.

Boeing's 4th generation 737 MAX's smallest variant, the MAX-7, is no different and has perfectly followed into the footsteps of its short-lived & commercially unsuccessful predecessors booking a mere 52 orders[7] (of the total 4000+ orders for the entire MAX series) as of June 2020, led by Boeing's die hard loyalist Southwest's 30 and WestJet's 22 aircraft orders, since the program's launch in 2011. Airbus A319neo, the smallest member of the A320neo family, too, has been no head turner either with the program's order book standing at a mere 84[8] aircraft orders, as of March 2020, of which 47 have come in 2020 alone. Compare that to the entire A320neo family's stellar performance with over 7000+ orders. That's exactly why the zone has been referred to colloquially as the narrow body segment's 'Bermuda Triangle'!

The C-Series from Bombardier, especially the smaller CS-100, was targeted at this very segment (with CS-300 covering the notch above) with Bombardier; maneuvering as the skilled marauding buccaneer looking to take a fresh dig at the segment; taking full advantage of the clean sheet, smaller sized, custom-built aircraft packing composites and a pair of next generation engines; which just couldn't have been matched either by Airbus or Boeing in operating economics with their existing portfolio and the history clearly being against squandering billions in fresh development in this segment.

The plan from Bombardier was sound; as reflected by the program's ascend under Airbus wing in its reincarnated avatar as the A220; but Bombardier's resource planning & execution were worthy of an outright 'F' grade and the C-Series, too, like others before disappeared in the very same 'Bermuda Triangle'!

Airbus, better watch out & keep tracking the A220-100's flight path ahead!

6. The 'Almost Forlorn' Wide Body Landscape

The wide body aircraft market has been a mere shadow of its former self impacted by onset of the global wave of protectionism underscored by the U.S.-China trade war & heightened U.S.-EU trade tensions, global macroeconomic environmental uncertainties, geopolitical tensions and the advent of the larger, extended variants of the A321neo family. The LR & XLR; have simply been biting into the lower end of the global wide body market pie having already captured a huge share of the middle of the market further compounding Boeing's woes & adding to the misery.

The LR & XLR's predatory ambitions & potential had been anticipated at the outset but this kind of ravenous instincts, monstrous gorging capacity & overall kill ratios had simply been unimaginable for anybody.

The A321neo program had a grounded start with 119 orders for the launch year 2011, unlike the A320neo which almost had a vertical take-off with 1000+ orders for the same year, showcasing its tremendous capabilities & future growth trajectory almost presciently. The A321neo's order book really took off from 2013 with the launch of A321LR in 2014 giving the program a sort of cryogenic boost with the A321neo booking 1000+ orders in the 3 year period spanning 2015-2017 followed by another surge in 2019 with 476 orders in that year alone reflecting the launch of the A321XLR in June 2019 at the Paris Air Show. By March 2020, the A321neo program had amassed 3,396 aircraft orders, as per Airbus orders & deliveries numbers[8].

With airlines ever looking to optimize profits to cost ratio by reducing the denominator, the longer A321 variants came in as a shot in the arm with their flexibility & capability to fly medium long haul, thin routes of 4,000+nmi range at unbeatable operating economics along with them being easy to fill (with 200 odd seats) which couldn't have been matched by the entry level wide bodies like the A330-800neo and the 787-8; which are not that easy to fill (400 seats in single class layout of A330-800 & 359 seats for the 787-8 in single class/242 in twin class at maximum capacity) along with a minimum 7,000-8,000nmi operating range.

Airlines have been shifting a part of the traditional long haul load on to these extended narrow bodies as of late under a revamped fleet strategy, which along with the larger macroeconomic factors, are being reflected in the order book of these programs.

The A330neo order book has been dominated by the larger sibling of the two, the -900, with the -800 accounting for a mere 14 aircraft orders, since the program's launch in 2014[8]. Boeing's 787 program, on the contrary, has fared much better on that count with the smallest 787-8 variant accounting roughly for just under a third of the cumulative 787 orders while the 787-9 has led the rally charging ahead relentlessly with almost 60% of total 787 orders.

The 787-8 has done well with its order intake on the ascent through 2011-2014 with the trend peaking in 2014 with 104 aircraft orders, followed by a quick fall off the pike from 2015 onwards towards average levels. The order intake for the 787-8[9] for the years 2011-2014 stood at 218 aircrafts while the orders tally for the years 2015-2019 declined by almost a third standing at 152 aircrafts with 2018 & 2019 proving to be the leanest years for the 787-8 since 2012 with order intake for these years contracting almost by 75% from the average level for the 2011-2019 period[9] of 41 aircrafts, as per Boeing sales numbers.

Boeing's concerns on the 787 program have been increasing over the recent years with the escalation of the U.S.-China trade war impacting orders for 787s originating from China with not even a single one of the Chinese airlines having placed any orders for wide-body aircrafts with Boeing since 2017.

Further, grounding of a faction of the global 787 fleet over problems with the Rolls Royce Trent 1000 engine (rapid parts wearing out); which powers the 787 as an engine option, have hit the 787 program & Rolls Royce hard over the recent years further compounding problems. Cancellation of a 22 aircraft order for 787s (worth $5.5 billion) by the Russian carrier Aeroflot in September 2019 came as another blow for Boeing.

The latest series of incoming salvo for Boeing has been the Norwegian's long range naval gun fire in form of the lawsuit over MAX grounding in July 2020 and 787's engine issues with Norwegian claiming over $1 billion in compensation, cancelling order for 100 737 MAX aircrafts & rescinding on the already delivered 737 MAXs & 787s.

The pressure on Boeing has been mounting on the 787 with the emergence & detection of quality issues on the program, in addition, to looming deferred accounting quantities, which, as the onboard excess baggage, have been undermining the program's thrust to weight ratio & profitability and in turn needing & necessitating higher production rates to offset.

With the order book position faltering and as a fallout of the COVID-19 outbreak, Boeing actually is going to lower production rates on the 787 to 7 aircrafts per month from 2022; a third successive rate cut following the earlier announced rate cut to 12 from late 2020 to be followed by a further ramp down to 10 per month from 2021; a number which has been revised & significantly scaled down further to 6 per month from 2021 post the second quarter financial results in July 2020.

Add the unprecedented impact of the COVID-19 to the overall mix and the overall picture gets even grimmer with wide bodies projected to be the last ones to recover given the severe, long term impact of pandemic on international air routes & traffic. The production rates on 777 & 777X are also being cut by Boeing to 3 aircrafts per month, down from 5 earlier with the 777 already in the ramp down phase while the 777X's order book has remained stagnant for a long time compounded by a year long delay on the program.

In an 'altered skyline' post pandemic, the sun has also set for the

'Queen of the Skies', the iconic 747, with Boeing announcing end of production run in 2022. Airbus, too, is facing a similar scenario with a dwindling A330neo order book & low production rates to be sustained.

However, Airbus is still slightly better positioned to deal with the post pandemic scenario with a narrow body segment led business structure as well as profile and a much stronger narrow body line-up than Boeing. Airbus, thus; with the A380 quietly fading away into the horizon annihilated by the pandemic; is likely to continue with the A330neo & the A350XWB as its lead twin wide body programs for the long term, despite clear pressures on the A330neo's order book, by biting the bullet through the post pandemic economic fallout somehow with low production rates to keep the line alive rather than leaning towards the kill switch given the A330neo's market positioning covering the top end of the middle of the market at a low capital cost and Airbus would not want to relent on that.

Airbus has throughout maintained that the A330neo's order woes are simply attributable to softness of the global wide body aircraft market and its misalignment with the fleet replacement cycle of the carriers. The order backlog positions for wide body aircraft programs across both Airbus (900+) & Boeing (1000+), in fact, have been tottering at just around the 1,000 aircraft units mark as of December 31, 2019 as against the peaks of 1400-1500 seen during the market's heydays.

Another key dilemma confronting Boeing, especially since COVID-19 outbreak, had been to carve out a future path ahead to manage & sustain low rate production on the 787 program, currently split across two final assembly lines located on either coast of the United States, namely, Everett, WA and Charleston, SC. The dual assembly structure, led by the establishment of the non-unionized Charleston facility, was created by

Boeing just a decade back as a bulwark against the negotiating heft of
unions at Everett.

However, Boeing, once again had been in a fix, stuck between the devil
and the deep blue sea post the pandemic's outbreak and was required to
make a difficult choice between the brutal quantitative logic of business
consolidation and the preservation instinct for its highly specialized &
arcane technical skill base engaged in airplane engineering & development
at Everett for decades with the problem for Boeing further compounded by
the capsizing of the NMA by the huge COVID-19 waves.

Boeing eventually has chosen the consolidation mode by going for
Charleston and abandoning the Everett line while also angering the
Washington state machinery. Boeing had found itself facing this kind of a
business dilemma in 2011 as well and that was on the 737 program when it
had to choose between opting for a bottom line-focused, business
economics led strategy approach by going for re-engining the 737 or to take
the time tested approach honoring safety as well as the sacrosanct laws of
aerospace engineering with a clean sheet approach and Boeing ultimately
ended up creating the MAX disaster.

Boeing, with a traditional wide body segment focus, thus, has a big task
ahead of itself with serious firefighting to do on a number of war fronts
while simultaneously also carrying out a major overhaul & reboot under the
boot followed by smoothening out of the blatant rough edges to reinvent
itself.

The jet streams that had been acting as the tailwinds at the onset of
1990s decade in form of the waves of globalization, which had enabled
Boeing to ride on them and build up its wide body empire, are now blowing
in exactly the opposite direction to become headwinds for the company, in

form of the waves of protectionism, favoring the narrow bodies and the middle of the market segments, a terrain which typically favors Airbus courtesy Boeing's consistent neglect of the segment coupled with a series of product strategy blunders in it. Boeing will have to quickly realign its strategic compass structurally in order to navigate a fundamentally altered market landscape in the post COVID-19 world effectively by reorienting itself rapidly.

But the million dollar question is whether Boeing is up to the massive challenge in the narrow body ring with an original playbook (hopefully) under a new 'Crew Chief' this time round who has a huge responsibility to adroitly marshal a beleaguered aerospace giant back onto its original flight line!

7. The 'WTO Ring'

The almost over a decade and a half long stalemate pertaining to aircraft launch subsidies (forming part of a larger set of trade tensions); underscored by the U.S. and the EU being at loggerheads over the contentious issue in the WTO ring; seems to finally be on the verge of a negotiated breakthrough.

Post the WTO verdict in favor of the U.S. in Q4 2019 and the WTO slamming both the U.S. & EU over faulting on the subsidies issue; the U.S. Trade Representative (USTR) had levied 10% tariffs on imports from Europe in October 2019, which subsequently had been raised to 15% in March 2020 hitting European exports to the U.S. really hard. The timing of the verdict has meant a relatively advantageous overall strategic position for the U.S. with the EU looking for a middle ground as the EU's counter salvo in the WTO still awaits ruling.

A statement from a European Commission representative in June 2020 captured the EU's take on the trade dispute. It said: "For the EU the priority remains finding a balanced negotiated solution to the aircraft subsidies[10]". Toeing that line, Airbus has come out with a subsidy concession plan to get the U.S. to back paddle on the tariffs part with the EU backing the bid for negotiated settlement backed by a credible deterrent in form of its own counter salvo of sanctions on the U.S., post the WTO verdict on EU case, which is likely by this year autumn.

An Airbus company press release; aimed at détente & signaling overtly towards peace overtures; was shot out, "the company plans to increase loan repayments to France & Spain with amendments to the Repayable Launch Investment (RLI) contracts through raise of interest rates applicable on loans taken for the development of its A350 program[11] [12]". Airbus further said that "the WTO has already ruled repayable launch investments a valid instrument for governments to partner with industry by sharing investment risks[11] [12]".

Post the announcement of subsidy concessions plan, geared towards rapprochement, Airbus further said: "The tariffs imposed by the USTR continue to harm all targeted industry sectors, including U.S. airlines, and have added to a difficult environment resulting from the COVID-19 crisis. This is why Airbus has decided to make a final step to remove the last contentious point and amend the French and Spanish contracts to what the WTO considers the appropriate interest rate and risk assessment benchmarks.

With this final move, Airbus considers itself in complete compliance with all WTO rulings[11] [12]". The U.S. had declared itself fully compliant with

WTO regulations in May 2020 following the annulment of tax sops focused on the aerospace industry.

Following a WTO verdict on the European case in October 2020, EU has just levied its own set of counter tariffs, amounting to 15% on $4 billion worth of U.S. imports, in early November 2020 amid hopes of improvement in trade relations & some headway towards resolution under the newly elected Biden administration.

The latest step taken by the EU might potentially open the so far fully jammed iron door slightly towards the possibility of a negotiated settlement in a post-COVID world, leaving the ball in the traditional U.S. 'hard court' now, seemingly marking the potential beginning of the end of one of the long standing trade disputes in commercial aviation history.

The economic aftermath of the virus seemingly might have just been the most undesired & the least wanted of the catalysts needed to reach the much required breakthrough to one of the most impregnable impasses in aviation history!

8. The 'Great Circle Route' to Aviation Safety's Next Pylon

Given that most air crashes in commercial aviation happen because of either Loss of Control In-Flight (LOC-I) with the uncontrolled flight crashing into the terrain or Controlled Flight into Terrain (CFIT) wherein, as per Boeing, "an airworthy aircraft under pilot control unintentionally and inadvertently goes into geographic features, which could be ground, rocky pikes or water bodies etc. with the crew unable to realize as to what is happening till it is too late"[12].

As per Boeing, CFIT has been one of the biggest causes of aviation deaths having already accounted for over 9,000 casualties by the mid-1990s since the onset of commercial jet operations[12]. CFIT has been classified as the second largest fatal accident category by IATA after LOC-I. There were about 47 CFIT accidents that took place between 2008-2017 of which 42 were fatal and caused 892 fatalities, as per IATA[13].

CFIT; emanating primarily from G induced Loss of Consciousness (G-LOC) or loss of spatial orientation; has been the biggest cause of crashes in military aviation as well given the excruciating, bone crushing impact rendered on the bodies of human pilots flying them. As per USAF, CFIT incidents account for almost 26% of aircraft losses & 75% of all pilot fatalities on the F-16 aircraft program alone[14]. Defense and especially Military aviation has been the sort of fountainhead of the aerospace's technological evolution traditionally with a large pool of technologies developed initially on the military side ultimately finding their way into commercial applications across the fence.

The list includes the fly by wire technology's insertion into the A320 family and usage of carbon composites in the airframes, which started with military aircrafts in the 1970s followed by their nano steps finding their way across the fence with a small cameo on the Airbus A300/310 in the same decade and ultimately reaching mainstream with their extensive utilization in commercial aviation by the early 21st century with the 787 & A350XWB programs almost having up to 50% utilization levels.

The saga also includes usage of Carbon Matrix Composite (CMCs) materials like Oxide Oxide® (OxOx) and proprietary surface coatings like Super Finish®, both developed originally by GE Aviation for military aircraft engines, into CFM International's latest LEAP engines. This is apart from numerous instances of turbofan engines developed originally for military aircrafts later evolving into commercial derivatives finding extensive commercial applications, with the trend led by the GE's F101 engine, which was developed originally for the B-1 Lancer Bomber program in 1974 and evolved into the commercially highly successful CFM56 a decade later. GE's CF6 engine, too, was derived from the TF39 turbofan engine which was developed originally for the USAF's C-5 Galaxy military transport aircraft program in the late 1960s.

As with all other complex problems, military aviation had been at work ardently to sort out CFIT issues as well. To address & prevent the CFIT situations, Lockheed Martin's 'Skunk Works', in collaboration with Air Force Research Laboratory and National Aeronautics and Space Administration (NASA), came up with the Automatic Ground Collision Avoidance System (A-GCAS) based on over 3 decades of tedious research work.

The A-GCAS system uses the triad of terrain mapping, geo-location & automation decision making algorithms & technologies to detect an impending ground collision for the aircraft and intervenes at first by alerting the pilot to take corrective action and in case it detects no change in aircraft's flight trajectory it automatically comes into action and during the penultimate seconds of an imminent crash (factoring in the high speeds, recovery rates & specific mission requirements for military jets) takes control of the aircraft temporarily by rolling up the aircraft's wings level and a strong high G pull-up maneuver to prevent the aircraft from crashing into the ground.

The system also has the option of Piloted Activated Recovery System (PARS), which could be activated by the Pilot consciously if they face spatial disorientation, vertigo or similar concerns etc. The system's immensely valuable capabilities & demonstrated success made Lockheed Martin Corporation the truly deserving recipient of the prestigious Robert J. Collier Trophy for the year 2018.

As per Lockheed Martin, the system was introduced on to the USAF's in-service fleet of F-16 aircrafts in 2014 and is in service on over 600 F-16 Block 40/50 aircrafts globally[14]. A-GCAS so far has been the sort of archangel, instrumental in saving around 9 F-16 aircrafts and 10 precious pilot lives[14] since the system's introduction.

Taking the game further to the next level, Lockheed Martin has also developed an Automatic Air Collision Avoidance System (A-ACAS) for the U.S. Government and has integrated it with the existing A-GCAS to have become the pioneer in the creation of the game changer Automatic Integrated Collision Avoidance System (A-ICAS). Lockheed Martin projects that the Automatic GCAS is likely to save 34 aircrafts, lives of 25 pilots and almost $2.3 billion over the next 15 years[14].

The purpose of including the A-GCAS in this discussion on commercial aviation is actually to uncover the holy grail of commercial aviation safety while it is still in the making. The U.S. Government & defense establishments have already shown the green flag for the integration of these cutting edge safety systems into the F-35 Lightning II aircraft program.

This could be followed by further development for integration into USAF's larger military aircrafts, like aerial tankers, commercially derived patrol aircrafts and bomber force which, currently is in consideration technically and could potentially open the adjacent door for their entry into mainstream commercial aviation market going forward as the aviation regulators have already zeroed in on to the game changer technology which

could literally catapult the industry's safety record onto an altogether
unthinkable trajectory like the turbofan engines have done with their overall
reliability levels over decades.

Currently, in the commercial aviation industry's existing inventory of
technologies, what comes closest to the A-GCAS are the Enhanced
Ground Proximity Warning Systems (EGPWS), which unfortunately seem
archaic & belonging to stone-age era when compared to the A-GCAS. The
EGPWSs almost reach the end of the road of their utility & application
with the sounding of mere non-specific aural warnings in the cockpit upon
detecting proximity to terrain and do not go beyond to the action level
intervention stage at all.

On the safety front of the multi-front war raging in commercial
aviation between Airbus & Boeing for decades; the A320 family has
trumped the Boeing's 737 aircraft program overall so far. The
comprehensive database compiled by AirSafe.com[17] shows in head-to-head
comparison that the overall fatal crash rates per million flights for the A320
family (ceo variants only as no crash has been reported so far for the neo
variants, thankfully!) at 0.09 has been substantially lower than the 0.24 for
the 737 program across generations with the 737 MAX (3.08); with its
almost rock bottom levels amongst existing, in-service aircraft programs;
taking the overall track record for the 737 way down.

However, comparing the A320 family data against the best 737
generation, the NG, unravels an entirely different picture. The 737NG's
fatal crash rate, at 0.07, is in fact marginally even better than the A320's
0.09! The statistics may just be indicating that the A320's onboard suite of
fly by wire & flight envelope protection technologies may not literally
translate into that direct connection & causative relationship with safety as

perceived with a definite scope for digging further deeper being there for sure for the inclined excavators!

The 737 program has progressively improved remarkably over generations given the crash rates for 737 Original (0.62), 737 Classic (0.14) and 737NG (0.07) with Boeing clearly having messed up terribly on the 737 MAX. The safest airplane program in the skies so far has surprisingly been a regional aircraft, the Embraer E170/190 family, with an overall crash rate of 0.03. Amongst the wide bodies, the safest program so far has been the 747-400 with a fatal crash rate of 0.06.

Coming back to A-GCAS; an installed, fully functioning A-GCAS system on-board fly-by-wire commercial aircrafts could be able to potentially save the hot headed, raging, out of control aircrafts from nose diving into the seas or terrains autonomously while also saving the ones in control from inadvertently flying into terrains as well. However, the fatal 737 MAX crashes may not have been prevented even with the A-GCAS, as the MAX, inheriting & still carrying the 1960s technological legacy doesn't have full fly-by-wire controls.

"If you know the enemy and know yourself, you need not fear the result of a hundred battles. If you know yourself but not the enemy, for every victory gained you will also suffer a defeat. If you know neither the enemy nor yourself, you will succumb in every battle." - Sun Tzu, Art of War

Chapter 14

FLIGHT PATHS LIKELY AHEAD…IN AN 'ALTERED SKYLINE'

"It is not the strongest of the species that survives, not the most intelligent…It is the one that is the most adaptable to change." – Charles Darwin

Crystal gazing is one activity which is one of the most difficult things to do in commercial aviation as the math is complex and the variables too many to corral and factor them in and project their interplay out into the horizon. The industry has defied many productions of downturns resiliently & turned them on their head while there have also been sudden downswings which were simply unpredictable by any stretch of human imagination, like the current COVID-19 pandemic & how it has decimated commercial aviation, for which there was simply no way anyone could have seen it coming let alone the notion of being prepared for it. However, looking at things as to how they stand, commercial aviation is in a tailspin for now and is likely to be there for another 3-5 years, through 2023 at the least, seemingly with the wide body aircraft segment likely to be the worst hit for the aircraft OEMs given international travel is likely to be the last one to be unshackled from the clutches of the virus.

Boeing faces a double whammy of blows with the supply side issues earlier now conjoined by the demand side ones as well now post pandemic. The gaping hole in Boeing's product portfolio in the middle of the market needs to be the first one to be plugged for Boeing; of course after getting the 737 MAX airborne again; for the company to have any meaningful possibility of combating the A321XLR's carnage in the segment in the absence of any worthy 757 successor.

Looking at the current scenario, the problem at hand and the airlines market trend the most optimum solution for Boeing to do this is likely to go the conventional way of initiating a clean sheet narrow body aircraft program, the Y1 leg of its project Yellowstone with the 757 specs, and developing it in two or three variants as a family of aircrafts, much like the A320 family.

The 757 replacement aircraft variant could then be stretched further to create the larger A321XLR killer while a shorter stretch would yield a potential 737 MAX replacement powered by next generation narrow body engines whenever required. If the 737 MAX does well post its return to the skies Boeing could take the call regarding succession based on emerging market developments & competitive scenario. The latest aircraft family could be packed with cutting edge & next generation technologies spearheaded by an all composite airframe and next generation engines.

The next evolution of engine technology is likely to come from Rolls Royce with the company's pursuit of the Ultrafan & Advance GTF engine programs likely to culminate into a next generation of engine family somewhere between 2025 & 2030. Pratt & Whitney & CFM International would surely be keeping an eye on that & doing their homework on matching the efficiency gains promised by Rolls Royce by stepping up their game on the PW1000GTF & LEAP engine technologies further. Boeing's Y1 is likely to be powered by that generation engine technology coming from any of the trio with Boeing unlikely to take a call on Y1 until the dust settles post the COVID-19 super sandstorm; unless there is a drastic issue with the 737 MAX's path to service return. Boeing is also likely to build in the capability to be able to eventually swap out the conventional turbofans powering the Y1 with a potential hybrid electric propulsion system or hydrogen based propulsion in future which is likely to be commercially available in initial iterations by late 2030s or early to mid 2040s.

The strategic advantage for Boeing with this aircraft family approach would be significant economic & scale advantages with the added ability to be able to cover multiple aircraft market segments (737, 757 and the just above 757 & right below 787 segment dominated by the A321XLR currently) with a single aircraft development program followed by derivatives strategy, unlike the original twin aisle NMA concept with single aisle operating economics, which was focused solely on the middle of the market as a niche program. This would be in line with the latest market trend being favored by the airlines of operating single aircrafts on long range thin routes, like the A321XLR, with the trend likely to get a further acceleration in the post COVID-19 world facing economic constraints.

An Airbus counterattack to this kind of a strategy would most likely be an outright further stretch of the A220 mounted with Rolls Royce Ultrafans to counter the 737 substitute variant to match performance levels. However, given Toulouse's traditional penchant for opening the game by fielding game changers, the A220's larger stretch as the A220-500 is likely to come proactively as the next game changer in the narrow body segment.

Airbus most likely would be looking at the operational play window starting right after the market recovers from the pandemic fall out, by 2023 or 2024 for the program launch, with the period extending up to 2035 as the power play phase for the A220-500 post which the age of carbon free commercial aircrafts is likely to dawn taking the game to another level altogether with the projected 30% enhanced fuel efficiency gains as against the traditional, current age fossil fuel guzzlers, as per existing Airbus game plan.

The launch of the A220-500, as the largest member of the small A220 family, makes sense for Airbus as a larger order book will push the production rate on the A220 program up and will thereby give the much needed traction to the profitability of the program by unlocking the A220's actual growth potential making the program find its ultimate 'True North'.

The post COVID-19 world also would be favorable for the A220 with smaller narrow bodies flying on domestic circuits are likely to be the first ones to stage a recovery and are likely to hold the line firmly. Further, the unleashing of the A220-500 would enable Airbus to effectively launch a classic pincer attack on the beleaguered 737 MAX and a subsequent crushing defeat for the already tainted MAX could then be very much be on the cards.

The pincer would in turn force Boeing to go for a new, clean sheet narrow body while simultaneously also spearheading a charge at the seemingly impregnable Airbus citadel right in the middle of the market without air cover. This could be achieved by Boeing by going either for revival of both NSA & NMA development in parallel or by developing a single clean sheet program doubling up as the middle of the market contender as well. Given the dire straits situation prevailing for commercial aviation post the COVID-19 outbreak the latter is likely to be Boeing's preferred choice (or may be the Hobson's choice) going forward.

Also, the A220-500 development will provide Airbus with the least cost option to keep its engineering technical skill base engaged meaningfully & intact till dust settles & the industry recovers from the pandemic sandstorm.

From a competitive standpoint, a 21st century scratch up, composite based A220 mainstream narrow body in the A320 & 737 MAX operating league would be a definite edge for Airbus, especially, with the 5% to 8% enhanced fuel efficiency gains it will be able to provide over the latest A320neo or the 737 MAX programs with their older airframes. And if that happens, Boeing; with nothing in its stable to be able to match the A220, will be even more hard pressed in the narrow body segment with its problems compounding further, especially with Boeing likely to be operating in the overhaul mode at least for the next half a decade.

Skeptics might talk about the potential cannibalization risks emerging for the A320neo program from the launch of the new, larger mustang; however, the solid, strategic positioning of the A320neo entrenched deeply at the core of the narrow body market as the proven workhorse, is likely to ensure its sustained 'tour de force' status for a long time to come.

Further, with the succession planning process for the A320 program having already been initiated, marking the beginning of the inevitable end, getting the sprinting mustang along might only give that extra steroid boost to the workhorse as the ultimate narrow body hunting pair capable of maneuvering & predating aggressively on MAX territory, looking to capture & take the narrow body market further away from Boeing. The bundling effects of the A320neo and the A220 pair on the Airbus order book have already been seen earlier.

For the potential 757-replacer & A321XLR competitor, Boeing would most likely have to go in for a clean sheet program to address the matter as it would need a composite airframe to match performance levels accordingly based on specs. However, all this is likely to play out; if it at all does happen this way; only in the latter half of the seemingly jinxed 2020s decade.

Airbus, thus, is likely to remain and is going to be on top of the narrow body league through at least the middle of the present century based on current, prevailing visibility levels from the cockpit as Airbus is not only playing to its narrow body segment focused strengths but is also expanding the narrow body market steadily wresting market share from Boeing, which had over the past 3 decades, placed all its bets complacently on the wide body market. Boeing will really have to play out of its skin with a major & much needed switch to proactive from reactive strategic operating mode in the narrow bodies to be able to catch up on the widening Airbus lead in the rapidly expanding narrow body segment.

The next generation of narrow body commercial airplanes, be it clean sheet designs or re-engined ones as determined by the prevailing market forces at the most appropriate time frames, are likely to appear from there-on only in the mid-2030s (with the pace of technology evolution accelerated & catalyzed almost by a decade by the COVID-19 outbreak), and they are likely to be mostly electrified having electric elements added to it in the propulsion system, be it hybrid electric or pure electric propulsion systems or hydrogen based propulsion, driven by the burgeoning sustainability challenges which are likely to into only one direction from here on and which is upwards and thus is likely to be a key defining force shaping the future of aviation.

However, recurrence of any COVID-19 like tsunamis or any other unseen force majeure again in the future is likely to kick the entire gamut of timelines, horizons, equations, projections & scenarios pertaining to the future simply out of the ballpark as effortlessly as it has happened this time round in 2020 without leaving any possibility & scope of being factored into potentially any kind of crystal ball gazing activity.

One such game changer move has already come with the French pushing the throttle to the max rpm mark to take their aviation game to the next level for that decisive edge, while the world grapples with the COVID-19 fallout, with an announcement in May 2020 to infuse billions of Euros into their domestic industry for next generational technologies.

This ambitious game plan is going to be spearheaded by the development of a clean sheet successor to the A320 by 2035 powered by a hydrogen based propulsion system becoming the world's first zero emissions aircraft in operation while also trifurcating the path towards sustainability with the addition of hydrogen to the biofuels and hybrid-electric pathways.

However, it will not be easy by any stretch of imagination given the significant evolutionary challenges to be overcome, especially, the volumetric power density problem of liquid hydrogen which is a mere 20% of the current in-use aviation fuels. And if the feisty French are able to do it, it will take them to the next level with a head start & a tremendous strategic advantage in the pole position with an unassailable lead and a significant competitive advantage that will be take time for the others to match.

Lastly, the long term trend to watch out for in commercial aviation (from a global perspective) is that airplanes; starting out as the ordinary steed of the royal in the post World War II world, became the ordinary steed of the above ordinary by the end of second half of the twentieth century, are moving towards becoming the ordinary steed of the ordinary in the first half of the twenty first century which is likely to be followed by their becoming the ordinary steed of the below ordinary in the second half of the twenty first century!

"Attention passengers, this is your Captain once again".

"With this we are approaching our destination and have already started the descent. Preparing for the landing now…"

"It's been a privilege to have you aboard on this journey. Hope you enjoyed the flight!"

Keep Flying. Godspeed!

Over & Out!

Bibliography

Chapter – 1

1. "From war to partner: Airbus and the CSeries" published by Leeham News and Analysis dated October 18, 2017

    https://leehamnews.com/2017/10/18/war-partner-airbus-cseries/

2. "CFM Tech Insertion bringing lower fuel burn, longer on-wing life", CFM International, November 15, 2009

    https://www.cfmaeroengines.com/press-articles/cfm-tech-insertion-bringing-lower-fuel-burn-longer-on-wing-life/

3. "Pictures: Airbus aims to thwart Boeing's narrowbody plans with upgraded 'A320 Enhanced'". Flight International. 20 June 2006

    https://www.flightglobal.com/pictures-airbus-aims-to-thwart-boeings-narrowbody-plans-with-upgraded-a320-enhanced/67962.article

4. "Don't assume A320, 737 RE Programs", Leeham News and Analysis, published March 30, 2010

    https://leehamnews.com/2010/03/29/dont-assume-a320-737-re-programs/?subscribe=success#blog_subscription-2

5. Singapore 2010: "Airbus targets early A320 re-engining decision, 2015 debut", published by FlightGlobal, Max Kingsley Jones, Februaru 4, 2010

    https://www.flightglobal.com/singapore-2010-airbus-targets-early-a320-re-engining-decision-2015-debut/91831.article

6. "Airbus set to launch A320neo", published by FlightGlobal, date December 01, 2010 by Jon Ostrower

https://www.flightglobal.com/airbus-set-to-launch-a320-neo/97169.article

Chapter – 2

1. "How A320 changed the world for commercial pilots", Flight International, published 20 February 2017, David Learmount

2. "Airbus at 50. Five things Airbus got right and five it didn't", Flight Global, By Murdo Morrison, published May 28, 2019
https://www.flightglobal.com/airbus-at-50/five-things-airbus-got-right-and-five-it-didnt/132315.article

3. "Transport News: Boeing Plans Jet." The New York Times, July 17, 1964. Retrieved: February 26, 2008
https://www.nytimes.com/1964/07/17/archives/transport-news-boeing-plans-jet-shorthaul-models-studied-appeal-set.html

4. "Boeing delivers its 5,000th 737", published by Seattle Pi, dated February 12, 2006
https://www.seattlepi.com/default/article/Boeing-delivers-its-5-000th-737-1195654.php

5. "Boeing 737 Model Summary – Orders and Deliveries", Boeing.com, Updated June 2020
http://active.boeing.com/commercial/orders/displaystandardreport.cfm?cboCurrentModel=737&optReportType=AllModels&cboAllModel=737&ViewReportF=View+Report

6. "Lift for Airbus from Pan Am" published by New York Times, September 14, 1984 by Lee A Daniels
https://www.nytimes.com/1984/09/14/business/lift-for-

airbus-from-pan-am.html

7. "Northwest, Pan Am have Airbus A320s on order with PM-France-Crash" AP News, June 27, 1988, https://apnews.com/25b8115e121f1ea98ae7be5edb997cfc

8. "Final Report concerning the accident which occurred on June 26, 1988 at Mulhouse-Habsheim to the Airbus A320, registered F-GFKC". November 29, 1989 https://reports.aviation-safety.net/1988/19880626-0_A320_F-GFKC.pdf

9. "The A320 Habsheim accident, An Airbus Industrie response to allegations made in television programmes and other media" https://www.aviation-accidents.net/report-download.php?id=90012

10. "Northwest, Pan Am have Airbus 320s on order with PM-France-Crash" Associated Press, June 27, 1988 https://apnews.com/25b8115e121f1ea98ae7be5edb997cfc

11. "Leahy reflects on 33 years at Airbus" published by Leeham News, November 28, 2017. https://leehamnews.com/2017/11/28/leahy-reflects-33-years-airbus/

12. "Brash Airbus sales chief Leahy bows out after reshaping the industry" published by Seattle Times, February 04, 2018 by Dominic Gates https://www.seattletimes.com/business/boeing-aerospace/brash-airbus-sales-chief-leahy-bows-out-after-reshaping-the-industry/

13. "Top Airbus Officials scoffed at Leahy's 50% market share goal" published by Leeham news, December 14, 2017 by Scott Hamilton. https://web.archive.org/web/20181129153740/https://leehamnews.com/2017/12/14/top-airbus-officials-scoffed-leahys-50-market-share-goal/

14. "Leahy to bow out from Airbus at year-end after final sales push", Reuters, October 11, 2017, By Tim Hepher https://www.reuters.com/article/us-airbus-leahy-idUSKBN1CG1QE

15. "Boeing must transform the way it builds planes" published by Fortune Classic,   March 08, 1993 https://fortune.com/1993/03/08/boeing-planes-manufacturing/

16. Airbus Cockpits, https://www.airbus.com/aircraft/passenger-aircraft/cockpits.html

17. "Composites in Aerospace Applications", Adam Quilter, AVIATIONPROS, October 01, 2004. https://www.aviationpros.com/engines-components/aircraft-airframe-accessories/article/10386441/composites-in-aerospace-applications

18. Airbus A320, Aerospace Technology, https://www.aerospace-technology.com/projects/a320/

19. "United Signs Airbus Deal in Blow to Boeing: Airlines: The Carrier says it will lease 50 of the European-built planes. Deal is worth an estimated $2.4 billion", Los Angeles Times, July 09, 1992, sourced from Reuters, https://www.latimes.com/archives/la-xpm-1992-07-09-fi-2575-story.html

20. "Almost cancelled, the Boeing 737 has endured 49 years", HeraldNet, Dan Catchpole, March 26, 2016 https://www.heraldnet.com/news/almost-cancelled-the-boeing-737-has-endured-49-years/

21. Boeing 737 orders & deliveries http://active.boeing.com/commercial/orders/displaystandardreport.cfm?cboCurrentModel=737&optReportType=AllModels&cboAllModel=737&ViewReportF=View+Report

22. McDonnell Douglas DC-9
https://en.wikipedia.org/wiki/McDonnell_Douglas_DC-9
https://www.boeing.com/history/products/dc-9.page

23. Global Aviation Data Management (GADM) accident database
https://www.iata.org/en/services/statistics/gadm/adx/

24. Boeing Aircraft Orders & Deliveries
http://www.boeing.com/commercial/#/orders-deliveries

25. Airbus Historical Aircraft Orders & Deliveries
http://www.airbus.com/fileadmin/media_gallery/files/reports_results_reviews/Summary_Historial_Orders_Deliveries_1974-2009.xls

Chapter 3

1. "Boeing-Japan airbus rival move", World Trade News, Financial Times (29273). p. 6. ISSN 0307-1766, By Donne, Michael (March 16, 1984).

2. "7J7: Boeing sets the pace". *Flight International*, By Moxon, Julian (October 26, 1985). Vol. 128 no. 3983. Seattle, Washington, USA. pp. 25–28.

3. "Deadline!: How premier organizations win the race against time". Author: Dan Carrison, ISBN 0814426778, 9780814426777, publisher Amacom 2003, pp. 170 –171, Chapter: 'Boeing's race to deliver the 777 wide-body'.

4. "For the love of flying". *Design News. March 4, 1996.*

5. *"The short, happy life of the prop-fan", Air & Space Magazine, Bill Sweetman, Published September 2005*

https://www.airspacemag.com/history-of-flight/the-short-happy-life-of-the-prop-fan-7856180/?all

6. "Boeing's Key Suppliers in Japan want an upgrade". Seattle Times. May 13, 2017, Dominic Gates

https://www.seattletimes.com/business/boeing-aerospace/boeings-key-suppliers-in-japan-want-an-upgrade/

7. "Boeing 7J7 design to be frozen in July". Interavia, Vol. 42 no. 1. pp. 23–26. Accessed via ACTUALITE Aéronautique

https://web.archive.org/web/20180131184535/http://avia.superforum.fr/t1200-boeing-7j7#25565

8. "Communication Standards take to the air", New Scientist, September 17, 1987, By Helen Gavaghan. Accessed via Google Books

https://books.google.fr/books?id=0uLgn2nma4EC&pg=PA44&lpg=PA44&dq=7J7+weight&source=bl&ots=hIVzTE2kOK&sig=Gvh99nQ-VAJCNNa9VLDvHm2rk9U&hl=fr&ei=0lylTcO4IoSV8QOxgL25Dw&sa=X&oi=book_result&ct=result#v=onepage&q=7J7%20weight&f=false

9. Airbus A320 Specifications

https://web.archive.org/web/20120124123133/http://www.airbus.com/aircraftfamilies/passengeraircraft/a320family/a320/specifications/

10. "Joint venture brings back propellers", New Scientist, March 13, 1986, By Helen Gavaghan, Accessed via Google Books

https://books.google.co.in/books?id=mYVNkaEJpz4C&pg=PA27&redir_esc=y#v=onepage&q&f=false

11. "7J7 the next new Boeing", Flying Magazine, May 1987, Page 37. Accessed via Google Books

https://books.google.co.in/books?id=Y3Y0Dn-Qqm0C&pg=PA37&redir_esc=y#v=onepage&q&f=false

Chapter 4

1. Spinoff, National Aeronautics and Space Administration (NASA), Office of Commercial Programs Technology Utilization Division, James Haggerty, August 1987, NTRS

https://ia801204.us.archive.org/5/items/NASA_NTRS_Archive_19880002195/NASA_NTRS_Archive_19880002195.pdf

2. "France backs UDF". Propulsion. Flight International. Vol. 130 no. 4042. Villaroche, France. December 20, 1986. p. 63. ISSN 0015-3710

3. Airbus technology Strategy for A320, Europe Report Science & Technology, JPRS-EST-86-027, October 06, 1986, FBIS, NTIS

https://ia903102.us.archive.org/7/items/DTIC_ADA337361/DTIC_ADA337361.pdf

4. *"The short, happy life of the prop-fan", Air & Space Magazine, Bill Sweetman, Published September 2005*

https://www.airspacemag.com/history-of-flight/the-short-happy-life-of-the-prop-fan-7856180/?all

Chapter 5

1. "Boeing raises estimate of jet market by 22%" Washington Post, March 07, 1989 by Polly Lane.

https://www.washingtonpost.com/archive/business/1989/03/07/boeing-raises-estimate-of-jet-market-by-22/46753a3f-baf1-47be-bd14-49ac7f76adc4/

2. "Boeing 737 Model Summary – Orders and Deliveries", Boeing.com, Updated June 2020

http://active.boeing.com/commercial/orders/displaystandardreport.cfm?cboCurrentModel=737&optReportType=AllModels&cboAllModel=737&ViewReportF=View+Report

3. "Analysis: Half-century milestone marks 737's enduring appeal", Flight Global, April 07, 2017, By Stephen Trimble

https://www.flightglobal.com/analysis-half-century-milestone-marks-737s-enduring-appeal/123565.article

4. "7E7 may decide Boeing's future here" Seattle Times, May 1, 2003 by Dominic Gates

https://archive.seattletimes.com/archive/?date=20030501&slug=7e7boeing01

5. "Will 787 program ever show an overall profit? Analysts grow more skeptical", Seattle Times, October 17, 2015 by Dominic Gates https://www.seattletimes.com/business/boeing-aerospace/will-787-program-ever-show-an-overall-profit-analysts-grow-more-skeptical/

Chapter 6

1. "THE 737 STORY: Smoke and mirrors obscure 737 and Airbus A320 replacement studies Published by FlighGlobal, February 07, 2006, Norris Guy

2. Boeing patent may provide glimpse into 737 replacement plan, Flight Global, September 23, 2010, Jon Ostrower
https://web.archive.org/web/20141006074327/http://www.flightglobal.com/blogs/flightblogger/2010/09/boeing_patent_may_provide_glim/

3. Boeing boss green-lights all-new next generation narrowbody, Flight Global, February 10, 2011, Jon Ostrower
https://www.flightglobal.com/boeing-boss-green-lights-all-new-next-generation-narrowbody/98345.article

4. "Boeing plans to develop new airplane to replace 737 MAX by 2030". By Reuters, November 05, 2014
https://www.reuters.com/article/us-boeing-ceo-737/boeing-plans-to-develop-new-airplane-to-replace-737-max-by-2030-idUSKBN0IP27320141105

Chapter 7

1. "Airbus outlines expected market impact of A320NEO", Published by Flight International, December 07, 2010, By Kerry Reals

2. "Boeing CEO: 'new airplane' to replace 737", Associate Press, February 10, 2011, by Freed, Joshua, NBC News. Associated Press.

3. *"Airbus with new order record at Paris Air Show 2011", Airbus SE Press Release, June 23, 2011*
http://www.defense-aerospace.com/article-view/release/126715/airbus-sets-new-paris-air-show-record-with-$72-billion-in-orders.html

4. "Jet Order by American is a Coup for Boeing's Rival", The New York Times, Clark Nicola, July 20, 2011

5. *"AMR Corporation announces largest aircraft order in history with Boeing & Airbus", American Airlines, July 20, 2011*
http://news.aa.com/news/news-details/2011/AMR-Corporation-Announces-Largest-Aircraft-Order-In-History-With-Boeing-And-Airbus-07202011/default.aspx

6. *"Boeing Launches 737 New Engine Family with Commitments for 496 airplanes from Five Airlines" Boeing's company press release, August 30, 2011.*
https://boeing.mediaroom.com/2011-08-30-Boeing-Launches-737-New-Engine-Family-with-Commitments-for-496-Airplanes-from-Five-Airlines

7. *"Lessons from the 737 MAX Crisis" by Facultatea de Inginerie Aerospatiala, by Octavian Thor Pleter, August 14, 2019*
http://www.aero.pub.ro/wordpress/index.php/en/2019/08/14/lessons-from-the-boeing-737-max-crisis/

8. *"What is the Boeing 737 MAX Maneuvering Characteristics Augmentation System?" The Air Current, November 13, 2018*
https://theaircurrent.com/aviation-safety/what-is-the-boeing-737-max-maneuvering-characteristics-augmentation-system-mcas-jt610/

9. *SKYbrary, Flight Safety Foundation,*
https://aviationsafetywiki.org/index.php/DC10

10. *'Boeing Builds Airplanes, McDonnell-Douglas Builds Character,' Pilots Recall As Delta Flies Final MD-88, MD-90 Trips" Ted Reed June 01, 2020, published by Forbes*

https://www.forbes.com/sites/tedreed/2020/06/01/as-
airlines-trim-fleets-delta-plans-final-md-88-and-md-90-trips-
for-tuesday/#46c742cc1d58

11. *"American Air Says Goodbye to MD-80 Jet After 36 Years" Published
by Bloomberg Quint, September 04, 2019 by Mary Schlangenstein*
https://www.bloomberg.com/news/articles/2019-09-
04/american-says-goodbye-to-md-80-jet-after-36-years-of-
love-hate

12. *"Crash Course: How Boeing's managerial revolution created the 737
MAX disaster" published by The New Republic , September 18, 2019,
by Maureen Tkacik*
https://newrepublic.com/article/154944/boeing-737-max-
investigation-indonesia-lion-air-ethiopian-airlines-managerial-
revolution

13. *"Did the Boeing egineers' strike of 2000 succeed in the long run?"
Published by knkx, By Ashley Gross, January 10, 2013*
https://www.knkx.org/post/did-boeing-engineers-strike-2000-
succeed-long-run

14. "The Long-Forgotten Flight That Sent Boeing Off-Course",
The Atlantic, November 20, 2019, By Jerry Useem
https://www.theatlantic.com/ideas/archive/2019/11/how-
boeing-lost-its-bearings/602188/

15. "Boeing strike another hit to economy", CNN Money,
September 12, 2008, By Chris Isidorc
https://money.cnn.com/2008/09/12/news/economy/boeing
_impact/index.htm

16. "The Role of Engineering in the Navy", by Admiral H.G.
Rickover, August 30, 1974
http://gmapalumni.org/chapomatic/extras/Rickover.htm

17. "Almost cancelled, the Boeing 737has endured 49 years",
HeraldNet, March 26, 2016, By Dan Catchpole
https://www.heraldnet.com/news/almost-cancelled-the-
boeing-737-has-endured-49-years/

Chapter 8

1. "Boeing to Buy McDonnell Douglas" The New York Times,
   December 16, 1996, By Brian Knowlton
   https://www.nytimes.com/1996/12/16/news/boeing-to-
   buy-mcdonnell-douglas.html

2. The Boeing Company, Annual Report for the year 2001.

3. "Boeing, Rockwell complete merger" Kitsap Sun, December
   06, 1996, by Sun news services cts.kitsapsun.com
   https://products.kitsapsun.com/archive/1996/12-
   06/356230_boeing__rockwell_complete_merge.html

4. "Boeing Must Transform the Way it Builds Planes" Fortune,
   March 8, 1993, By Shawn Tully
   https://fortune.com/1993/03/08/boeing-planes-
   manufacturing/

5. Tichy, N. (1989). GE's Crotonville: A Staging Ground for
   Corporate Revolution. *The Academy of Management Executive
   (1987-1989), 3*(2), 99-106. Retrieved August 12, 2020, from
   www.jstor.org/stable/4164880

6. "Winging it at Boeing& #8217;s Leadership Center",
   Workforce.com, September 29, 2000, By Caroline Cole
   https://www.workforce.com/news/winging-it-at-boeings-
   leadership-center

7. "The Long-Forgotten Flight That Sent Boeing Off-Course",
   The Atlantic, November 20, 2019, By Jerry Useem
   https://www.theatlantic.com/ideas/archive/2019/11/how-
   boeing-lost-its-bearings/602188/

8. "Boeing: What Really Happened" By Bloomberg Week Online, December 15, 2003

https://www.bloomberg.com/news/articles/2003-12-14/boeing-what-really-happened

9. "Boeing Reports Loss For 1997", The Washington Post, January 28, 1998, By Tim Smart

https://www.washingtonpost.com/archive/business/1998/01/28/boeing-reports-loss-for-1997/03650606-c6f7-4899-9925-c08abf2df4d6/

10. The Boeing Company, Annual Report for 1997.

11. "Boeing probe intensifies over secret Lockheed papers", The Seattle Times, January 09, 2005, By David Bowermaster.

https://www.seattletimes.com/business/boeing-probe-intensifies-over-secret-lockheed-papers/

12. "Airbus agrees to settle corruption probes with U.S., UK and France", CNBC, Jan 28, 2020

https://www.cnbc.com/2020/01/28/airbus-agrees-to-settle-corruption-probes-with-us-uk-and-france.html

13. "Final Committee Report: The Design, Development & Certification of the Boeing 737 MAX", the House Committee on Transportation & Infrastructure, September 2020.

https://transportation.house.gov/imo/media/doc/2020.09.15%20FINAL%20737%20MAX%20Report%20for%20Public%20Release.pdf

14. Airbus Commercial Aircraft Results Summary, 1989-2019.xls

https://www.google.com/url?sa=t&rct=j&q=&esrc=s&source=web&cd=&cad=rja&uact=8&ved=2ahUKEwjb9o7sybHsAhVGAXIKHdZ7A_oQFjACegQIAxAC&url=https%3A%2F

%2Fwww.airbus.com%2Fcontent%2Fdam%2Fcorporate-
topics%2Fpublications%2Fbackgrounders%2Fairbus-
commercial-aircraft-results-summery-1989-
2019.xls&usg=AOvVaw1yC1ry8OmW2qAs7_vt7hov

15. "Airbus breaks into BA with huge A320 order", FlightGlobal,
September 02, 1998, Max Kingsley-Jones/Toulouse
https://www.flightglobal.com/airbus-breaks-into-ba-with-
huge-a320-order/22699.article

15. "100 Airbuses for Northwest", Chicago Tribune, October 02,
1986, Carol Jouzaitis
https://www.chicagotribune.com/news/ct-xpm-1986-10-02-
8603140154-story.html

16. "United Signs Airbus Deal in Blow to Boeing: Airlines: The
Carrier says it will lease 50 of the European-built planes. Deal
is worth an estimated $2.4 billion", Los Angeles Times, July 09,
1992, sourced from Reuters,
https://www.latimes.com/archives/la-xpm-1992-07-09-fi-
2575-story.html

17. "Boeing 737 Model Summary – Orders and Deliveries",
Boeing.com, Updated June 2020
http://active.boeing.com/commercial/orders/displaystandard
report.cfm?cboCurrentModel=737&optReportType=AllModel
s&cboAllModel=737&ViewReportF=View+Report

18. Airbus Historical Aircraft Orders & Deliveries
http://www.airbus.com/fileadmin/media_gallery/files/reports
_results_reviews/Summary_Historial_Orders_Deliveries_1974
-2009.xls

Chapter 9

1. "Bombardier officially begins construction of Belfast CSeries site" Published by FlightGlobal, Lori Ranson, 17 November 2009. https://www.flightglobal.com/bombardier-officially-begins-construction-of-belfast-cseries-site/90428.article

2. "AirAsia boss confirms talks for 100 CSeries CS300". Farnborough: Published by Flightglobal, Govindasmy, Siva dated July 12, 2012

3. "Bombardier Weighs Third CSeries Jet Model". *The Wall Street Journal*. New York. Published May 21, 2015. Jon Ostrower

4. "CSeries Aircraft Program Making Excellent Progress Towards First Flight", *Bombardier Press release, 7 March 2013*

5. "Transport Canada Certifies Pratt & Whitney PurePower® PW1500G Engine for Bombardier CSeries Aircraft". *Bombardier Press release. February 20, 2013* https://www.bombardier.com/en/media/newsList/details.41744-transport-canada-certifies-pratt-whitney-purepower-r-pw1500g-engine-for-bombardier-cseries-aircraft.bombardiercom.html

6. "Bombardier Completes CSeries Aircraft Ground Vibration Tests and Final Software Upgrades In Preparation for First Flight". Bombardier Press release, *June 26, 2013* https://www.bombardier.com/en/media/newsList/details.bombardier-completescseriesaircraftgroundvibrationtestsandfinals.bombardiercom.html

7. "Bombardier's CSeries Aircraft Completes Historic First
Flight". Bombardier Press Release. September 16, 2013
https://www.bombardier.com/en/media/newsList/details.b
ombardier-aerospace20130916cseriesfirstflight.html

8. "Bombardier C Series: record orders in 2016 as both variants
finally enter service" Centre for Aviation, CAPA, December
08, 2016.
https://centreforaviation.com/analysis/reports/bombardier-
c-series-record-orders-in-2016-as-both-variants-finally-enter-
service-317615

9. "Boeing gives United a Smoking Deal on 737s to block
Bombardier from gaining Traction" Forbes, March 08, 2016.
By Scott Hamilton.
https://www.forbes.com/sites/scotthamilton5/2016/03/08/
united-boeing-and-the-competitors/#6d9380f230da

10. "Bombardier-Delta deal can put Boeing out of business,
company claims", Leeham News & Analysis, May 25, 2017,
https://leehamnews.com/2017/05/25/bombardier-delta-
deal-can-put-boeing-business-company-claims/

11. "Airbus and Bombardier Announce C-Series Partnership"
Airbus press release, October 16, 2017
https://www.airbus.com/newsroom/press-
releases/en/2017/10/airbus-bombardier-cseries-
agreement.html

12. "From war to partner: Airbus and the CSeries" published by
Leeham News and Analysis dated October 18, 2017
https://leehamnews.com/2017/10/18/war-partner-airbus-
cseries/

13. "Boeing and Embraer to Establish Strategic Aerospace Partnership to Accelerate Global Aerospace Growth", Boeing press release, July 05, 2018 https://boeing.mediaroom.com/2018-07-05-Boeing-and-Embraer-to-Establish-Strategic-Aerospace-Partnership-to-Accelerate-Global-Aerospace-Growth

14. "Boeing terminates Agreement to Establish Joint Ventures with Embraer" Boeing company press release, April 25, 2020 https://boeing.mediaroom.com/2020-04-25-Boeing-Terminates-Agreement-to-Establish-Joint-Ventures-with-Embraer

15. "Boeing ends deal, angering Brazilian jet maker Embraer", Defense News, April 26, 2020, By The Associated Press https://www.defensenews.com/air/2020/04/26/boeing-ends-deal-angering-brazilian-jet-maker-embraer/

Chapter 10

1. Boeing 757 Commercial Transport, Historical Snapshot, Boeing https://www.boeing.com/history/products/757.page

2. A321ceo, the most efficient single aisle jetliner, key figures, Airbus https://www.airbus.com/aircraft/passenger-aircraft/a320-family/a321ceo.html

3. "Boeing close to decision on 757-200X launch" Flightglobal, October 13, 1999, By Max Kingsley https://www.flightglobal.com/boeing-close-to-decision-on-757-200x-launch/28845.article

4. "World Airline Census 2018". *Flightglobal, August 21, 2018.*

5. "Exclusive: Airbus launches "A321neoLR" long range to replace 757-200W" Leeham News & Analysis https://leehamnews.com/2014/10/21/exclusive-airbus-launches-a321neolr-long-range-to-replace-757-200w/

6. Airbus Orders & Deliveries, December 31, 2017.

7. "Airbus launches longest range single-aisle airliner: the A321XLR" Airbus company press release, June 17, 2019 https://www.airbus.com/newsroom/press-releases/en/2019/06/airbus-launches-longest-range-singleaisle-airliner-the-a321xlr.html

8. "PARIS: Airbus details design changes of A321XLR", Flightglobal, 17 June 2019, Kaminski-Morrow, David

9. "Air Canada would study Airbus, Boeing for transatlantic narrowbody needs." Reuters, January 15, 2019. Tim Hepher. https://www.reuters.com/article/us-airbus-canada-aircanada/air-canada-would-study-airbus-boeing-for-transatlantic-narrowbody-needs-idUSKCN1P82FT

10. Tell Me Why: "A lighter aircraft and the capability to fly farther" — Robert Isom, American Airlines press release, June 19, 2019 http://news.aa.com/news/news-details/2019/Tell-Me-Why-A-lighter-aircraft-and-the-capability-to-fly-farther--Robert-Isom/

11. "Rolls Royce Snecma Olympus", IHS Jane's, https://www.janes.com/transport/news/jae/jae000725_1_n.shtml

12. "United Studying Replacements for its Transatlantic 757s", AIN Online, March 11, 2015, by Gregory Polek

https://www.ainonline.com/aviation-news/air-transport/2015-03-11/united-studying-replacements-its-transatlantic-757s

13. "Boeing looking closely at an all-new jet for 2025" The Seattle Times, June 15, 2015, by Dominic Gates https://www.seattletimes.com/business/boeing-aerospace/boeing-looking-closely-at-an-all-new-jet-for-2025/

14. "Where is the middle of the market", Flight Ascend Consultancy, May 12, 2017, Chris Seymour, https://www.flightglobal.com/analysis-where-is-the-middle-of-the-market/123876.article

15. "General Electric unconvinced about demand for Boeing's 797" The Irish Times, July 16, 2018, https://www.irishtimes.com/business/transport-and-tourism/general-electric-unconvinced-about-demand-for-boeing-s-new-797-1.3567152

16. "Boeing wrestles with options for new midsize jet" The Seattle Times, March 02, 2016, By Dominic Gates. https://www.seattletimes.com/business/boeing-aerospace/boeing-wrestles-with-options-for-new-mid-sized-jet/

17. "NMA market sector is small, Airbus' Leahy says" Leeham News & Analysis, January 03, 2018 https://leehamnews.com/2018/01/03/nma-market-sector-small-airbus-leahy-says/

18. "Forget the NMA, go after A321-say ex-Airbus exec", Leeham News & Analysis, January 29, 2019,

https://leehamnews.com/2019/01/29/forget-the-nma-go-after-a321-say-ex-airbus-exec/

19. "Rolls-Royce sees NMA "addressable" market as 4,000-5,000, same as Boeing", Leeham News & Analysis, January 22, 2019 https://leehamnews.com/2019/01/22/rolls-royces-sees-nma-addressable-market-as-4000-5000-same-as-boeing/

20. "Pontifications: Doubts continue over Boeing NMA launch", Leeham News & Analysis, March 04, 2019, Scott Hamilton, https://leehamnews.com/2019/03/04/pontifications-doubts-continue-over-boeing-nma-launch/

Chapter 11

1. Aircraft Accident Investigation Bureau Preliminary Report. Ethiopian Civil Aviation Authority, Ministry of Transport (Ethiopia). March 2019. http://www.ecaa.gov.et/Home/wp-content/uploads/2019/07/Preliminary-Report-B737-800MAX-ET-AVJ.pdf

2. "Indonesian Plane Crash Adds to Country's Troubling Safety Record" The New York Times, October 28, 2018, By Muktita Suhartono and Hannah Beech https://www.nytimes.com/2018/10/28/world/asia/indonesia-lion-air-plane-crash.html

3. "EU lifts ban on all Indonesian airlines" June 14, 2018 https://eeas.europa.eu/delegations/indonesia/46513/eu-lifts-air-ban-all-indonesian-airlines_en

4. "Continued Airworthiness Notification to the International Community", FAA. March 11, 2019

    https://www.faa.gov/news/updates/?newsId=93206

    https://www.faa.gov/news/updates/media/CAN_2019_03.pdf

5. Lion Air 737 MAX Final Accident Report Cites AOA Sensor, MCAS Among Multitude of Contributing Factors

    https://www.aviationtoday.com/2019/10/28/lion-air-737-max-final-accident-report-cites-aoa-sensor-mcas-as-contributing-factors/

6. "Assumptions Used in the Safety Assessment Process and the Effects of Multiple Alerts and Indications on Pilot Performance", Safety Recommendation Report, National Transportation Safety Board, Washington D.C., October 29, 2018/ March 10, 2019 https://trid.trb.org/view/1658639

7. "Exclusive: Boeing kept FAA in the dark on key 737 MAX design changes – U.S. IG report, Reuters, July 01, 2020, David Shepardson, Eric M. Johnson, Tracy Rucinski

    https://in.reuters.com/article/boeing-737 MAX/exclusive-boeing-kept-faa-in-the-dark-on-key-737-max-design-changes-u-s-ig-report-idINL1N2E7344

8. Congressional Panel says Boeing has 'culture of concealment', The Associated Press, David Koeing, March 06, 2020.

    https://www.bostonglobe.com/2020/03/06/business/congressional-panel-says-boeing-has-culture-concealment/

9. "Boeing Employees Mocked FAA and 'Clowns' who Designed 737 MAX", The New York Times, January 09, 2020, By Natalie Kitroeff

https://www.nytimes.com/2020/01/09/business/boeing-
737-messages.html

10. "Boeing whistleblower alleges systemic problems with 737
MAX", The Seattle Times, June 18, 2020, by Dominic Gates
https://www.seattletimes.com/business/boeing-
aerospace/boeing-whistleblower-alleges-systemic-problems-
with-737-max/

11. *SKYbrary, Flight Safety Foundation,*
https://aviationsafetywiki.org/index.php/DC10

12. "Pilots Criticize Boeing, Saying 737 MAX 'Should Never
Have Been Approved', NPR, June 19, 2019, David Schaper,
https://www.npr.org/2019/06/19/734248714/pilots-
criticize-boeing-saying-737-max-should-never-have-been-
approved

13. "Final Committee Report: The Design, Development &
Certification of the Boeing 737 MAX", The House
Committee on Transportation & Infrastructure, September
2020
https://transportation.house.gov/imo/media/doc/2020.09.
15%20FINAL%20737%20MAX%20Report%20for%20Publ
ic%20Release.pdf

14. "House panel blasts Boeing and FAA over fatal 737 MAX
crashes", CBS News, September 16, 2020, By Kris Van
Cleave, https://www.cbsnews.com/news/boeing-737-max-
faa-house-report-condemns/

15. "Fitch Downgrades Boeing to 'BBB-'; Outlook Negative",
Fitch Ratings, October 29, 2020,
https://www.fitchratings.com/research/corporate-

finance/fitch-downgrades-boeing-to-bbb-outlook-negative-29-10-2020

16. "American Airlines Eyes Dec. 29 Takeoff for Boeing 737 MAX", The Street, By Adam Smith, October 18, 2020 https://www.thestreet.com/investing/american-airlines-eyes-boeing-737-max-takeoff-by-dec-29

17. Aircraft Certification Reform and Accountability Act, https://republicans-transportation.house.gov/uploadedfiles/aircraft_certification_reform_and_accountability_act_-_section_by_section.pdf

Chapter 12

1. Airbus Aircraft orders & deliveries https://www.airbus.com/aircraft/market/orders-deliveries.html

2. Boeing Annual Report 2019, p.29

3. The Boeing 737 Technical Site, http://www.b737.org.uk/sales.htm, Chris Brady

4. "Airbus offers new fuel saving engine options for A320 Family" Airbus press release, December 01, 2010 https://www.airbus.com/newsroom/press-releases/en/2010/12/airbus-offers-new-fuel-saving-engine-options-for-a320-family.html

5. "Airbus reports Full-Year (FY) 2019 results, delivers on guidance", https://www.airbus.com/newsroom/press-

releases/en/2020/02/airbus-reports-full-year-2019-results.html

6. "Boeing Reports Fourth-Quarter Results"

https://boeing.mediaroom.com/2020-01-29-Boeing-Reports-Fourth-Quarter-Results

7. "A320's order total overtakes 737's as Max crisis persists" FlightGlobal, November 15, 2019, By David Kaminski-Morrow

https://www.flightglobal.com/orders-and-deliveries/a320s-order-total-overtakes-737s-as-max-crisis-persists/135347.article

Chapter 13

1. "How big was the global climate strike? 4 million people, activists estimate". Vox, September 22, 2019, By Eliza Barclay and Brian Resnick https://www.vox.com/energy-and-environment/2019/9/20/20876143/climate-strike-2019-september-20-crowd-estimate

2. "Air travel is a huge contributor to climate change. A new global movement wants you to be ashamed to fly". Vox, November 30, 2019. By Umair Irfan. https://www.vox.com/the-highlight/2019/7/25/8881364/greta-thunberg-climate-change-flying-airline

3. "What is CORSIA and how does it work?" ICA Environment https://www.icao.int/environmental-protection/Pages/A39_CORSIA_FAQ2.aspx

4. "E-Fan X: A giant leap towards zero-emission flight", Airbus, https://www.airbus.com/innovation/zero-emission/electric-flight/e-fan-x.html

5. AS2, Aerion Supersonic, https://www.aerionsupersonic.com/as2/

6. "Boom Preps for Rollout of XB-1 Supersonic Demonstrator", AIN Online, July 18, 2020, By Kerry Lynch https://www.ainonline.com/aviation-news/air-transport/2020-07-18/boom-preps-rollout-xb-1-supersonic-demonstrator

8. The Boeing 737 Technical Site, http://www.b737.org.uk/sales.htm, Chris Brady

9. Airbus Aircraft orders & deliveries https://www.airbus.com/aircraft/market/orders-deliveries.html

10. Boeing Aircraft Orders & Deliveries http://www.boeing.com/commercial/?cm_re=March_2015--Roadblock-_-Orders+%26+Deliveries/#/orders-deliveries

11. "Brussels warns new US tariff threat over Airbus will harm both sides" Financial Times, June 24, 2020, BY Jim Brunsden and Aime Williams https://www.ft.com/content/7cabefbe-5a58-4f96-96c4-e18718d17e44

12. "Airbus takes final step to end long-standing WTO dispute and U.S. tariffs" Airbus press release, July 24, 2020

https://www.airbus.com/newsroom/press-releases/en/2020/07/airbus-takes-final-step-to-end-longstanding-wto-dispute-and-us-tariffs.html

13. Boeing: Commercial Airplanes – Jetliner Safety - Industry's Role in Aviation Safety, June 29, 2011
https://boeing.mediaroom.com/1997-02-20-Boeing-Training-Aid-Addresses-Leading-Accident-Cause

14. Global Aviation Data Management (GADM) accident database
https://www.iata.org/en/services/statistics/gadm/adx/

15. "Saving the Good Guys with Auto GCAS Technology" Lockheed Martin Corporation
https://www.lockheedmartin.com/en-us/products/autogcas.html

16. "Jet builder Aerion expects to fly silent supersonic planes by 2024, unlocking a $40 billion market", CNBC, Michael Sheetz, Jan 16, 2020
https://www.cnbc.com/2020/01/16/aerions-as2-silent-supersonic-jet-aims-to-fly-by-2024.html

17. Plane crash rates by model, Air Safe.com
http://www.airsafe.com/events/models/rate_mod.htm

18. "Wicker Releases Committee's Investigation Report", U.S. Senate Committee on Commerce, Science & Transportation
https://www.commerce.senate.gov/2020/12/wicker-releases-committee-s-faa-investigation-report

List of Illustrations & Images

Chapter 3

Image 1: McDonnell Douglas MD-91 propfan airliner project desktop model at Western Museum of Flight, (Converted to monochrome)

Image Credits: Extrapolaris, Wikimedia Commons,

CC-A-SA 4.0 International

Chapter 4

Image 1: Model of the Unducted Fan (UDF) Engine on Display at the Safran Museum

Image Credits: Duch, Wikimedia Commons,

CC-A–SA 4.0 International

Image 2: The GE36 on a McDonnell Douglas MD-80 demonstrator at the 1988 Farnborough Air Show.

Image Credits: Andrew Thomas, Wikimedia Commons, CC BY-SA 2.0

Chapter 5

Image 1: A 737-200 belonging to the Aerolíneas Argentinas

Image Credits: Markus Hening, Wikimedia Commons, Public

Domain (PD)

Image 2: An Air North 737-201

Image Credits: Makaristos, Wikimedia Commons,

Public Domain (PD)

Image 3: A CFM56-3 Engine Mounted on a Japan TransOcean Airlines 737-400

Image Credits: Wikimedia Commons, CC BY-SA 3.0

Image 4: A CFM56-5 Engine on an A320

Image Credits: Curimedia, Wikimedia Commons,

CC BY 2.0

Chapter 7:

Image 5: A Boeing 737-400's CFM56-3 engine, with its recognizable non-circular "hamster pouch" inlet

Image Credits: Davidelit, Wikimedia Commons,

Public Domain

Image 6: CFM56-7B of a 737-800(NG series)

Image Credits: Trainler, Altair78, Wikimedia Commons,

CC BY 3.0

Image 7: CFM-56 turbofan engine on a Boeing 737 aircraft

Image Credits: Aviacsa, Wikimedia Commons,

Public Domain (PD)

Image 8: Mounting position of LEAP-1B on a Norwegian Air International's Boeing 737 MAX

Image Credits: Edward Russell, Wikimedia Commons,

C-C-A 2.0 Generic License

Illustration 1: CFM56-7B vs. LEAP-1B – Engine Length & Fan Diameter

Illustration 2: CFM56-7B vs. LEAP-1B – Dry Weight

Illustration 3: CFM56-7B vs. LEAP-1B – Max Take-Off Thrust

Illustration 4: LEAP-1A vs. 1B vs. 1C Variants - Engine Length & Fan Diameter

Illustration 5: LEAP-1A vs. 1B vs. 1C Variants - Max Take-Off Thrust

Illustration 6: LEAP-1A vs. 1B vs. 1C Variants – Max Height & Width

Data Sources: Illustrations 1-6 Above

1. *CFM International* https://www.safran-aircraft-engines.com/commercial-engines/single-aisle-commercial-jets/cfm56/cfm56-7b

2. *EASA Type certificate data sheet for LEAP-1A & -1C engines* https://web.archive.org/web/20181013014334/https://www.easa.europa.eu/sites/default/files/dfu/EASA%20E110%20TCDS%20Issue%207%20LEAP-1A-1C.pdf

3. *LEAP-1B Engines* https://www.easa.europa.eu/sites/default/files/dfu/EASA%20E115%20TCDS%20Issue%207%20LEAP-1B.pdf

Chapter 9

1. "Bombardier officially begins construction of Belfast CSeries site" Published by FlightGlobal, Lori Ranson, 17 November 2009. https://www.flightglobal.com/bombardier-officially-begins-construction-of-belfast-cseries-site/90428.article

2. "AirAsia boss confirms talks for 100 CSeries CS300". Farnborough: Published by Flightglobal, Govindasmy, Siva dated July 12, 2012

3. "Bombardier Weighs Third CSeries Jet Model". *The Wall Street Journal*. New York. Published May 21, 2015. Jon Ostrower

4. "CSeries Aircraft Program Making Excellent Progress Towards First Flight", *Bombardier Press release, 7 March 2013*

5. "Transport Canada Certifies Pratt & Whitney PurePower®
    PW1500G Engine for Bombardier CSeries Aircraft".
    Bombardier Press release. February 20, 2013
    https://www.bombardier.com/en/media/newsList/details.41
    744-transport-canada-certifies-pratt-whitney-purepower-r-
    pw1500g-engine-for-bombardier-cseries-
    aircraft.bombardiercom.html

6. "Bombardier Completes CSeries Aircraft Ground Vibration
    Tests and Final Software Upgrades In Preparation for First
    Flight". Bombardier Press release, *June 26, 2013*
    https://www.bombardier.com/en/media/newsList/details.b
    ombardier-
    completescseriesaircraftgroundvibrationtestsandfinals.bombar
    diercom.html

7. "Bombardier's CSeries Aircraft Completes Historic First
    Flight". Bombardier Press Release. September 16, 2013
    https://www.bombardier.com/en/media/newsList/details.b
    ombardier-aerospace20130916cseriesfirstflight.html

8. "Bombardier C Series: record orders in 2016 as both variants
    finally enter service" Centre for Aviation, CAPA, December
    08, 2016.
    https://centreforaviation.com/analysis/reports/bombardier-
    c-series-record-orders-in-2016-as-both-variants-finally-enter-
    service-317615

9. "Boeing gives United a Smoking Deal on 737s to block
    Bombardier from gaining Traction" Forbes, March 08, 2016.
    By Scott Hamilton.

https://www.forbes.com/sites/scotthamilton5/2016/03/08/united-boeing-and-the-competitors/#6d9380f230da

10. "Bombardier-Delta deal can put Boeing out of business, company claims", Leeham News & Analysis, May 25, 2017, https://leehamnews.com/2017/05/25/bombardier-delta-deal-can-put-boeing-business-company-claims/

11. "Airbus and Bombardier Announce C-Series Partnership" Airbus press release, October 16, 2017 https://www.airbus.com/newsroom/press-releases/en/2017/10/airbus-bombardier-cseries-agreement.html

12. "From war to partner: Airbus and the CSeries" published by Leeham News and Analysis dated October 18, 2017 https://leehamnews.com/2017/10/18/war-partner-airbus-cseries/

13. "Boeing and Embraer to Establish Strategic Aerospace Partnership to Accelerate Global Aerospace Growth", Boeing press release, July 05, 2018 https://boeing.mediaroom.com/2018-07-05-Boeing-and-Embraer-to-Establish-Strategic-Aerospace-Partnership-to-Accelerate-Global-Aerospace-Growth

14. "Boeing terminates Agreement to Establish Joint Ventures with Embraer" Boeing company press release, April 25, 2020 https://boeing.mediaroom.com/2020-04-25-Boeing-Terminates-Agreement-to-Establish-Joint-Ventures-with-Embraer

15. "Boeing ends deal, angering Brazilian jet maker Embraer", Defense News, April 26, 2020, By The Associated Press

https://www.defensenews.com/air/2020/04/26/boeing-ends-deal-angering-brazilian-jet-maker-embraer/

Chapter 10

1. Boeing 757 Commercial Transport, Historical Snapshot, Boeing https://www.boeing.com/history/products/757.page

2. A321ceo, the most efficient single aisle jetliner, key figures, Airbus https://www.airbus.com/aircraft/passenger-aircraft/a320-family/a321ceo.html

3. "Boeing close to decision on 757-200X launch" Flightglobal, October 13, 1999, By Max Kingsley https://www.flightglobal.com/boeing-close-to-decision-on-757-200x-launch/28845.article

4. "World Airline Census 2018". *Flightglobal, August 21, 2018.*

5. "Exclusive: Airbus launches "A321neoLR" long range to replace 757-200W" Leeham News & Analysis https://leehamnews.com/2014/10/21/exclusive-airbus-launches-a321neolr-long-range-to-replace-757-200w/

6. Airbus Orders & Deliveries, December 31, 2017.

7. "Airbus launches longest range single-aisle airliner: the A321XLR" Airbus company press release, June 17, 2019 https://www.airbus.com/newsroom/press-releases/en/2019/06/airbus-launches-longest-range-singleaisle-airliner-the-a321xlr.html

8. "PARIS: Airbus details design changes of A321XLR", Flightglobal, 17 June 2019, Kaminski-Morrow, David

9. "Air Canada would study Airbus, Boeing for transatlantic narrowbody needs." Reuters, January 15, 2019. Tim Hepher. https://www.reuters.com/article/us-airbus-canada-aircanada/air-canada-would-study-airbus-boeing-for-transatlantic-narrowbody-needs-idUSKCN1P82FT

10. Tell Me Why: "A lighter aircraft and the capability to fly farther" — Robert Isom, American Airlines press release, June 19, 2019 http://news.aa.com/news/news-details/2019/Tell-Me-Why-A-lighter-aircraft-and-the-capability-to-fly-farther--Robert-Isom/

11. "Rolls Royce Snecma Olympus", IHS Jane's, https://www.janes.com/transport/news/jae/jae000725_1_n.shtml

12. "United Studying Replacements for its Transatlantic 757s", AIN Online, March 11, 2015, by Gregory Polek https://www.ainonline.com/aviation-news/air-transport/2015-03-11/united-studying-replacements-its-transatlantic-757s

13. "Boeing looking closely at an all-new jet for 2025" The Seattle Times, June 15, 2015, by Dominic Gates https://www.seattletimes.com/business/boeing-aerospace/boeing-looking-closely-at-an-all-new-jet-for-2025/

14. "Where is the middle of the market", Flight Ascend Consultancy, May 12, 2017, Chris Seymour, https://www.flightglobal.com/analysis-where-is-the-middle-of-the-market/123876.article

15. "General Electric unconvinced about demand for Boeing's 797" The Irish Times, July 16, 2018,

https://www.irishtimes.com/business/transport-and-tourism/general-electric-unconvinced-about-demand-for-boeing-s-new-797-1.3567152

16. "Boeing wrestles with options for new midsize jet" The Seattle Times, March 02, 2016, By Dominic Gates.
https://www.seattletimes.com/business/boeing-aerospace/boeing-wrestles-with-options-for-new-mid-sized-jet/

17. "NMA market sector is small, Airbus' Leahy says" Leeham News & Analysis, January 03, 2018
https://leehamnews.com/2018/01/03/nma-market-sector-small-airbus-leahy-says/

18. "Forget the NMA, go after A321-say ex-Airbus exec", Leeham News & Analysis, January 29, 2019,
https://leehamnews.com/2019/01/29/forget-the-nma-go-after-a321-say-ex-airbus-exec/

19. "Rolls-Royce sees NMA "addressable" market as 4,000-5,000, same as Boeing", Leeham News & Analysis, January 22, 2019
https://leehamnews.com/2019/01/22/rolls-royces-sees-nma-addressable-market-as-4000-5000-same-as-boeing/

20. "Pontifications: Doubts continue over Boeing NMA launch", Leeham News & Analysis, March 04, 2019, Scott Hamilton,
https://leehamnews.com/2019/03/04/pontifications-doubts-continue-over-boeing-nma-launch/

Chapter 11

1. Aircraft Accident Investigation Bureau Preliminary Report.
Ethiopian Civil Aviation Authority, Ministry of Transport
(Ethiopia). March 2019. http://www.ecaa.gov.et/Home/wp-
content/uploads/2019/07/Preliminary-Report-B737-
800MAX-ET-AVJ.pdf

2. "Indonesian Plane Crash Adds to Country's Troubling Safety
Record" The New York Times, October 28, 2018, By Muktita
Suhartono and Hannah Beech
https://www.nytimes.com/2018/10/28/world/asia/indonesi
a-lion-air-plane-crash.html

3. "EU lifts ban on all Indonesian airlines" June 14, 2018
https://eeas.europa.eu/delegations/indonesia/46513/eu-lifts-
air-ban-all-indonesian-airlines_en

4. "Continued Airworthiness Notification to the International
Community", FAA. March 11, 2019
https://www.faa.gov/news/updates/?newsId=93206
https://www.faa.gov/news/updates/media/CAN_2019_03.p
df

5. Lion Air 737 MAX Final Accident Report Cites AOA Sensor,
MCAS Among Multitude of Contributing Factors
https://www.aviationtoday.com/2019/10/28/lion-air-737-
max-final-accident-report-cites-aoa-sensor-mcas-as-
contributing-factors/

6. "Assumptions Used in the Safety Assessment Process and the
Effects of Multiple Alerts and Indications on Pilot
Performance", Safety Recommendation Report, National

Transportation Safety Board, Washington D.C., October 29, 2018/ March 10, 2019 https://trid.trb.org/view/1658639

7. "Exclusive: Boeing kept FAA in the dark on key 737 MAX design changes – U.S. IG report, Reuters, July 01, 2020, David Shepardson, Eric M. Johnson, Tracy Rucinski https://in.reuters.com/article/boeing-737 MAX/exclusive-boeing-kept-faa-in-the-dark-on-key-737-max-design-changes-u-s-ig-report-idINL1N2E7344

8. Congressional Panel says Boeing has 'culture of concealment', The Associated Press, David Koeing, March 06, 2020. https://www.bostonglobe.com/2020/03/06/business/congressional-panel-says-boeing-has-culture-concealment/

9. "Boeing Employees Mocked FAA and 'Clowns' who Designed 737 MAX", The New York Times, January 09, 2020, By Natalie Kitroeff https://www.nytimes.com/2020/01/09/business/boeing-737-messages.html

10. "Boeing whistleblower alleges systemic problems with 737 MAX", The Seattle Times, June 18, 2020, by Dominic Gates https://www.seattletimes.com/business/boeing-aerospace/boeing-whistleblower-alleges-systemic-problems-with-737-max/

11. *SKYbrary, Flight Safety Foundation,* https://aviationsafetywiki.org/index.php/DC10

12. "Pilots Criticize Boeing, Saying 737 MAX 'Should Never Have Been Approved', NPR, June 19, 2019, David Schaper, https://www.npr.org/2019/06/19/734248714/pilots-

criticize-boeing-saying-737-max-should-never-have-been-approved

13. "Final Committee Report: The Design, Development & Certification of the Boeing 737 MAX", The House Committee on Transportation & Infrastructure, September 2020
https://transportation.house.gov/imo/media/doc/2020.09.15%20FINAL%20737%20MAX%20Report%20for%20Public%20Release.pdf

14. "House panel blasts Boeing and FAA over fatal 737 MAX crashes", CBS News, September 16, 2020, By Kris Van Cleave, https://www.cbsnews.com/news/boeing-737-max-faa-house-report-condemns/

15. "Fitch Downgrades Boeing to 'BBB-'; Outlook Negative", Fitch Ratings, October 29, 2020,
https://www.fitchratings.com/research/corporate-finance/fitch-downgrades-boeing-to-bbb-outlook-negative-29-10-2020

16. "American Airlines Eyes Dec. 29 Takeoff for Boeing 737 MAX", The Street, By Adam Smith, October 18, 2020
https://www.thestreet.com/investing/american-airlines-eyes-boeing-737-max-takeoff-by-dec-29

17. Aircraft Certification Reform and Accountability Act,
https://republicans-transportation.house.gov/uploadedfiles/aircraft_certification_reform_and_accountability_act_-_section_by_section.pdf

Chapter 12

1. Airbus Aircraft orders & deliveries https://www.airbus.com/aircraft/market/orders-deliveries.html

2. Boeing Annual Report 2019, p.29

3. The Boeing 737 Technical Site, http://www.b737.org.uk/sales.htm, Chris Brady

4. "Airbus offers new fuel saving engine options for A320 Family" Airbus press release, December 01, 2010 https://www.airbus.com/newsroom/press-releases/en/2010/12/airbus-offers-new-fuel-saving-engine-options-for-a320-family.html

5. "Airbus reports Full-Year (FY) 2019 results, delivers on guidance", https://www.airbus.com/newsroom/press-releases/en/2020/02/airbus-reports-full-year-2019-results.html

6. "Boeing Reports Fourth-Quarter Results" https://boeing.mediaroom.com/2020-01-29-Boeing-Reports-Fourth-Quarter-Results

7. "A320's order total overtakes 737's as Max crisis persists" FlightGlobal, November 15, 2019, By David Kaminski-Morrow https://www.flightglobal.com/orders-and-deliveries/a320s-order-total-overtakes-737s-as-max-crisis-persists/135347.article

Chapter 13

5. "How big was the global climate strike? 4 million people, activists estimate". Vox, September 22, 2019, By Eliza Barclay and Brian Resnick https://www.vox.com/energy-and-environment/2019/9/20/20876143/climate-strike-2019-september-20-crowd-estimate

6. "Air travel is a huge contributor to climate change. A new global movement wants you to be ashamed to fly". Vox, November 30, 2019. By Umair Irfan. https://www.vox.com/the-highlight/2019/7/25/8881364/greta-thunberg-climate-change-flying-airline

7. "What is CORSIA and how does it work?" ICA Environment https://www.icao.int/environmental-protection/Pages/A39_CORSIA_FAQ2.aspx

8. "E-Fan X: A giant leap towards zero-emission flight", Airbus, https://www.airbus.com/innovation/zero-emission/electric-flight/e-fan-x.html

5. AS2, Aerion Supersonic, https://www.aerionsupersonic.com/as2/

6. "Boom Preps for Rollout of XB-1 Supersonic Demonstrator", AIN Online, July 18, 2020, By Kerry Lynch https://www.ainonline.com/aviation-news/air-transport/2020-07-18/boom-preps-rollout-xb-1-supersonic-demonstrator

19. The Boeing 737 Technical Site, http://www.b737.org.uk/sales.htm, Chris Brady

20. Airbus Aircraft orders & deliveries

https://www.airbus.com/aircraft/market/orders-deliveries.html

21. Boeing Aircraft Orders & Deliveries

http://www.boeing.com/commercial/?cm_re=March_2015-_-Roadblock-_-Orders+%26+Deliveries/#/orders-deliveries

22. "Brussels warns new US tariff threat over Airbus will harm both sides" Financial Times, June 24, 2020, BY Jim Brunsden and Aime Williams

https://www.ft.com/content/7cabefbe-5a58-4f96-96c4-e18718d17e44

23. "Airbus takes final step to end long-standing WTO dispute and U.S. tariffs" Airbus press release, July 24, 2020

https://www.airbus.com/newsroom/press-releases/en/2020/07/airbus-takes_final-step-to-end-longstanding-wto-dispute-and-us-tariffs.html

24. Boeing: Commercial Airplanes – Jetliner Safety - Industry's Role in Aviation Safety, June 29, 2011

https://boeing.mediaroom.com/1997-02-20-Boeing-Training-Aid-Addresses-Leading-Accident-Cause

25. Global Aviation Data Management (GADM) accident database

https://www.iata.org/en/services/statistics/gadm/adx/

26. "Saving the Good Guys with Auto GCAS Technology" Lockheed Martin Corporation

https://www.lockheedmartin.com/en-us/products/autogcas.html

27. "Jet builder Aerion expects to fly silent supersonic planes by
2024, unlocking a $40 billion market", CNBC, Michael
Sheetz, Jan 16, 2020

https://www.cnbc.com/2020/01/16/aerions-as2-silent-
supersonic-jet-aims-to-fly-by-2024.html

28. Plane crash rates by model, Air Safe.com

http://www.airsafe.com/events/models/rate_mod.htm

29. "Wicker Releases Committee's Investigation Report", U.S.
Senate Committee on Commerce, Science & Transportation

https://www.commerce.senate.gov/2020/12/wicker-
releases-committee-s-faa-investigation-report

List of Illustrations & Images

Chapter 10

Image 10: A321XLR Infographic, Airbus Press Kits, Image
Credits: Airbus SE

Chapter 12

Illustration 1: Airbus vs. Boeing – Narrow Body Aircraft
Orders & Deliveries

Illustration 2: Airbus - Order Intake Split by Narrow Body
Aircraft Programs

Illustration 3: Boeing 737 Program – Order Intake Split by Generations

Illustration 4: A320 vs. 737 – Order Backlog - in Aircraft Units

Illustration 5: Airbus vs. Boeing – Cumulative Order Intake Split across Wide & Narrow Body Segments

Data Sources for Illustrations 1-5 Above

1. Airbus Aircraft orders & deliveries
   https://www.airbus.com/aircraft/market/orders-deliveries.html
2. Boeing Annual Report 2019, pp. 27,29
3. The Boeing 737 Technical Site,
   http://www.b737.org.uk/sales.htm, Chris Brady

ABOUT THE AUTHOR

Rajat Narang is the Co-Founder and Partner of a niche research firm pivoted on the Global Aerospace & Defense Industry for over a decade now and has authored over 1500+ syndicated research reports (across industries & sectors) providing strategic analysis on industries, sectors, players & markets with the end users being senior executives of most leading Global Commercial, Regional & Business Aviation OEMs (apart from defense primes), led by Airbus, Boeing, Bombardier, Embraer, Gulfstream, Dassault, Textron Aviation and engine OEMs as well as T1 Suppliers, including, GE Aviation, Rolls Royce, Pratt & Whitney, Safran and Spirit Aerosystems.

His educational background includes a Masters in Business Administration (MBA) - International Business with Business Strategy at the core followed by a Master's in Political Science with specialization in International Relations. Bitten early by the A&D bug while growing up, he has been actively following the industry for almost 3 decades now.